EAI/Springer Innovations in Communication and Computing

Series Editor

Imrich Chlamtac, European Alliance for Innovation, Ghent, Belgium

Editor's Note

The impact of information technologies is creating a new world yet not fully understood. The extent and speed of economic, life style and social changes already perceived in everyday life is hard to estimate without understanding the technological driving forces behind it. This series presents contributed volumes featuring the latest research and development in the various information engineering technologies that play a key role in this process.

The range of topics, focusing primarily on communications and computing engineering include, but are not limited to, wireless networks; mobile communication; design and learning; gaming; interaction; e-health and pervasive healthcare; energy management; smart grids; internet of things; cognitive radio networks; computation; cloud computing; ubiquitous connectivity, and in mode general smart living, smart cities, Internet of Things and more. The series publishes a combination of expanded papers selected from hosted and sponsored European Alliance for Innovation (EAI) conferences that present cutting edge, global research as well as provide new perspectives on traditional related engineering fields. This content, complemented with open calls for contribution of book titles and individual chapters, together maintain Springer's and EAI's high standards of academic excellence. The audience for the books consists of researchers, industry professionals, advanced level students as well as practitioners in related fields of activity include information and communication specialists, security experts, economists, urban planners, doctors, and in general representatives in all those walks of life affected ad contributing to the information revolution.

About EAI

EAI is a grassroots member organization initiated through cooperation between businesses, public, private and government organizations to address the global challenges of Europe's future competitiveness and link the European Research community with its counterparts around the globe. EAI reaches out to hundreds of thousands of individual subscribers on all continents and collaborates with an institutional member base including Fortune 500 companies, government organizations, and educational institutions, provide a free research and innovation platform.

Through its open free membership model EAI promotes a new research and innovation culture based on collaboration, connectivity and recognition of excellence by community.

More information about this series at http://www.springer.com/series/15427

Anandakumar Haldorai
Arulmurugan Ramu • Syed Abdul Rehman Khan
Editors

Business Intelligence for Enterprise Internet of Things

 Springer

Editors
Anandakumar Haldorai (iD)
Computer Science and Engineering
Sri Eshwar College Engineering
Coimbatore, Tamil Nadu, India

Arulmurugan Ramu (iD)
Presidency University
Rajanakunte, Yelahanka, India

Syed Abdul Rehman Khan
Tsinghua University
Beijing, China

ISSN 2522-8595 ISSN 2522-8609 (electronic)
EAI/Springer Innovations in Communication and Computing
ISBN 978-3-030-44409-9 ISBN 978-3-030-44407-5 (eBook)
https://doi.org/10.1007/978-3-030-44407-5

This Springer imprint is published by the registered company Springer Nature Switzerland AG
The registered company address is: Gewerbestrasse 11, 6330 Cham, Switzerland

Preface

The Internet of Things (IoT) has become an important research domain as enterprise applications, systems, infrastructures, and their applications have shown their potential in recent years. IoT is an enterprise bound for significant growth, and it will have a major impact on the lives of consumers and professionals around the world. It will enable enterprise industry to be a multitrillion dollar industry by 2025, including enterprise manufacturing, enterprise transportation, enterprise smart market, enterprise utilities, enterprise healthcare, etc. It will also change the way we think about producer and consumer networks. The expectations of IoT and its relevant products in this new era are quite high. Instead of smartness alone, consumers of IoT products and services would like to see IoT technologies bring about more intelligent systems and environments.

This book, *Business Intelligence for Enterprise Internet of Things*, presents the most recent challenges and developments in enterprise intelligence with the objective of promoting awareness and best practices for the real world. It aims to present new directions for further research and technology improvements in this important area. Its chapters include IoT enterprise system architecture, IoT-enabling enterprise technologies, and IoT enterprise services and applications, for example, enterprise on demands, market impacts, and its implications on smart technologies, big data enterprise management, and future enterprise Internet design for various IoT use cases, such as share markets, healthcare, smart cities, smart environments, smart communications, and smart homes.

This book also covers ideas, methods, algorithms, and tools for the in-depth study of performance and reliability of business intelligence for enterprise Internet of Things. The scope of business intelligence is to explore and present numerous research contributions relating to the field of neural network computing, business specifications, evolutionary computation, enterprise modeling and simulation, web intelligence, healthcare informatics, social relationship, energy, and end-to-end security in enterprise-aware management system in enterprise Internet of Things.

In this book, we present techniques and detailed perspectives of business intelligence for enterprise Internet of Things that can be used in overcoming and solving complex tasks in enterprise system. This book is based on various research horizons and contributions focusing on IoT enterprise system challenges over:

- Development of innovative enterprise architecture for the Internet of Things
- Enterprise IoT modeling: supervised, unsupervised, and reinforcement learning
- The Internet of Things evolutionary computation, enterprise modeling, and simulation
- Development of new IoT technologies for business intelligence and large-scale enterprise analysis
- Uncertainty modeling in big data analytics for IoT
- Providing solutions to pressing problems across areas including connected and autonomous vehicles, automation, healthcare, and enterprise security using the Internet of Things
- The management of enterprise in mobile transparent computing for the Internet of Things
- Bridge developments in artificial intelligence to real enterprise applications in collaboration with IoT partners
- New generation of scientists to address the skills shortage in these areas and increase competitiveness
- Applications and services for enterprise systems such as complex systems, multi-agent systems, game theory, and statistics
- Advanced future perspective in enterprise for the Internet of Things

This book opens the door for authors toward current research in enterprise Internet of Things systems for business intelligence.

We would like to thank Ms. Mary E. James, Senior Editor, Applied Sciences, Springer, and Ms. Eliska Vlckova, Managing Editor, European Alliance for Innovation (EAI), for their great support.

We anticipate that this book will open new entrance for further research and technology improvements. All the chapters provide a complete overview of business intelligence for enterprise Internet of Things. This book will be handy for academicians, research scholars, and graduate students in engineering discipline.

Coimbatore, Tamil Nadu, India   Anandakumar Haldorai
Rajanakunte, Yelahanka, India   Arulmurugan Ramu
Beijing, China   Syed Abdul Rehman Khan

Contents

About the Authors

Anandakumar Haldorai is Associate Professor and Research Head in the Department of Computer Science and Engineering, Sri Eshwar College of Engineering, Coimbatore, Tamil Nadu, India. He has received his Master's in Software Engineering from PSG College of Technology, Coimbatore, and his PhD in Information and Communication Engineering from PSG College of Technology under Anna University, Chennai. His research areas include Big Data, Cognitive Radio Networks, Mobile Communications, and Networking Protocols. He has authored more than 82 research papers in reputed international journals and IEEE conferences and 9 books and several book chapters in reputed publishers such as Springer and IGI. He is Editor in Chief of Keai-Elsevier *IJIN* and Inderscience *IJISC* and Guest Editor of several journals with Elsevier, Springer, Inderscience, etc. Also, he served as a Reviewer for IEEE, IET, Springer, Inderscience, and Elsevier journals and has been the General Chair, Session Chair, and Panelist in several conferences. He is Senior Member of IEEE, IET, and ACM and Fellow Member of EAI research group.

Arulmurugan Ramu is a Professor at Presidency University, Bangalore, India. His research focuses on the automatic interpretation of images and related problems in machine learning and optimization. His main research interest is in vision, particularly high-level visual recognition. He has authored more than 35 papers in major computer vision and machine learning conferences and journals. He completed his PhD in Information and Communication Engineering and his MTech and BTech in Information Technology all from Anna University of Technology, Chennai. He has guided several PhD research scholars under the area of Image Processing using machine learning. He is an Associate Editor of Inderscience *IJISC* journal. He was awarded Best Young Faculty Award 2018 and nominated for Best Young Researcher Award (Male) by International Academic and Research Excellence Awards (IARE-2019).

Syed Abdul Rehman Khan is an expert in Supply Chain and Logistics Management. He achieved his CSCP (Certified Supply Chain Professional) Certificate from the USA and successfully completed his PhD in China. Since 2018, he has been affiliated with Tsinghua University as a Postdoctoral Researcher. He has more than 12 years' core experience of supply chain and logistics at industry and academic levels. He has attended several international conferences and has been invited as Keynote Speaker in different countries. He has published more than 155 scientific research papers in different well-renowned international peer-reviewed journals and conferences. He has authored four books related to the sustainability in supply chain and business operations. He is a Regular Contributor to conferences and workshops around the world. In the last 2 years, he has won five different national-/provincial-level research projects. In addition, he has achieved scientific innovation awards three times consecutively by the Education Department of Shaanxi Provincial Government, China. Also, he holds memberships in the following well-renowned institutions and supply chain bodies/associations: APCIS, USA; Production and Operations Management Society, India; Council of Supply Chain Management Professionals, USA; Supply Chain Association of Pakistan; and Global Supply Chain Council, China.

Chapter 1
Internet of Things (IoTs) Evolutionary Computation, Enterprise Modelling and Simulation

A. Haldorai ⓘ, **A. Ramu** ⓘ, **and M. Suriya** ⓘ

1.1 Introduction

The Internet of Things (IoTs), for a few decades, has constituted a lot of vital topics concerning the future state of industries. Information and Communication Technology (ICT) is applicable in small telecommunication devices, which are affordable and regarded more effectively in terms of processing to accessing the internet. Moreover, Big Data technology was founded to allow businesses to store massive amounts of information, and evaluate the incoming streams of data with refined algorithms in actual time. As such, the evolution of the Internet of Things (IoTs) has allowed companies to create useful remedies for various case scenarios in various domains. In that regard, the IoTs can be regarded as the shelter terminology for various disciplines, which have already been considered in organizational automations. IoTs also enhances the integration of technological disciplines such as the automation of sensing information with enterprise resource organization data.

Despite the fact that IoTs technologies are being promoted extensively, just a few of the disciplines utilize cases which have to be implemented in organizational cases. Contrary to that, there are about a hundred of IoTs technologies and platforms, which have not been exploited yet. Moreover, there are various initiatives that explain various technological standards and their respective developments from establishments. The oversupply of the technological vendors enhances the implementation of integrated tech solutions. For instance, the research on development of

A. Haldorai (✉)
Sri Eshwar College of Engineering, Coimbatore, Tamil Nadu, India

A. Ramu
Presidency University, Bangalore, India

M. Suriya
KPR Institute of Engineering and Technology, Coimbatore, Tamil Nadu, India

© Springer Nature Switzerland AG 2020
A. Haldorai et al. (eds.), *Business Intelligence for Enterprise Internet of Things*,
EAI/Springer Innovations in Communication and Computing,
https://doi.org/10.1007/978-3-030-44407-5_1

IoTs suggest that interoperability, security and internet connectivity represent approximately three vital concerns in implementing various Internet Technology (IT) solutions. For the industries and the developers that have implemented the IoTs remedies, the performances initiated by them have now been considered based on connectivity. As such, this reflects our evaluation, which means that the rate of performance is considered based on architecture and developments created by users within a network cycle.

1.2 Background Analysis of Internet of Things (IoTs) for Modern Manufacturing

In this part, the connection of novel manufacturing paradigms and Internet Technology (IT) are discussed. The necessities of the organization models for effective adaption of IoTs are also considered. Many novel infrastructures for organizations are considered first. The discussions indicate that IoTs is planned effectively based on the architecture of the manufacturing industries. Consistently, every system element in the ES requires a data unit, which enhances the implementation of decisions on the elementary behaviours of the obtained dataset. A part from that, the acquisition of information, information transfer and the process of decision-making are fundamental functions that have to be considered in each model. In reference to the Reliability Theory, the Internet of Things is capable of proposing novel remedies that enhance scheduling, planning and the control of manufacturing organization at any level.

1.2.1 IoTs Robustness Evaluated Based on the Reliability Theory

Based on the aspect of scalability, the idea of reliability and robustness is considered. Thus, the IoTs shall be composed on a lot of electronic appliances whereby a lot of them are difficult to reconfigure. The idea of replacement is different from reconfiguration on tablet or desktop computers, which necessitate regular implementation of software updates to enhance modern upgrades. These updates require CPU power, memory and disk space. As mentioned by the Metcalfe's Laws, more novel interconnectivity between systems and subsystems enhances the degree of system probability that is concerned with the evaluation of system failures. The capability of IoTs to function for a long time irrespective of the condition of software and hardware is crucial to achieving user trust and acceptance. The bits errors can possibly lead to uncontrollable issues in massive networks, which might possibly characterize the IoTs. OR methods like the Reliability Theory will potentially be used to forecast on IoTs reliability and robustness. As such, there was need to formulate an approach based on the Reliability theory for the present subsystem networks, which are dependent on applications and perception layers. The total reliability RT is calculated using the below formula.

$$R_T = R_1 \times R_5 \times \left(1 - \pi \frac{4}{k=2} \left(1 - R_k \right) \right) \qquad (1.1)$$

In the Eq. (1.1), R1 and R5 represent the relative forms of reliabilities of application and perception layers, whereas R2, R3 and R4 show the reliability dependent on the internet, satellite networks and mobile networks. This framework can therefore be updated and expanded in accommodating the remaining IoTs features and characteristics of novel IoTs devices in operation. As such, the RT can potentially be applied onto the remaining IoTs features like data reliability. This application initiated the introduction of an application known as Fog Computing onto the IoTs whereas the researcher in [1] evaluated reliability of the methodology through the combination necessities of cloud and grid based on actuators and sensors in the IoTs. Resultantly, it was evident that reliability subsystems can be controlled. The IoTs software reliability in the IoTs can potentially be another important concern because software reliability may be termed as a special concern in RT. The reliability of hardware can be considered based on material or component failure, preventing a subsystem to perform as intended earlier. The reliability of software in the IoTs is considerably difficult to evaluate because software can potentially provide unanticipated findings for a lot of reasons like unutilized information retrieved from another networking appliance, which has not been noticed in the designing phase. The obsolescence of the entrenched software in the IoTs subsystems might not be maintained, altering reliability.

The researchers in [2] recommend a vital technique in which IoTs can be applied in business to enhance productivity. Nonetheless, the application of IoTs may differ for various data subsystems when it comes to providing high-resolution information in actual time. As such, this diminishes the overall costs of moving information from the actual world into the virtual world; for instance, the RFID tags can potentially eliminate costly manual stocks in the process of computing information into the computer. Based on the research, there are various means of applying IoTs in business to boost productivity. These include proximity triggers such as self-checking in libraries, automated sensor triggers like the networking smoke determiners and automatic product securities. As such, this indicates that IoTs represents the quantum leap-forward retrieved through the internet with the capability of acting as an element that governs the entire discipline for managing complex systems and organizations. Moreover, System Dynamics (SD) has also been implemented in organizational investment based on new technologies, particularly the incorporation of internet in China and India. Principally, organizational investment in the IoTs can potentially be modelled in the same manner. As such the researchers in [3] acknowledge that SD can potentially be represented to quantify the 'soft' variables that are fundamental in the process of incorporation of both technical and social aspects.

The major segment in SD defines the results evident in the behaviour of subsystems from interactions in feedback loops. Actually, different standardized feedback archetypes and looks have been noticed, including the 'fails and fixes' and the 'burdens and shifts'. In that manner, the dominant element of the internet incorporation was, over the years, argued to be the Contagion factor. With this effect or factor, innovators begin the process of adoption before developing the process of communication. This contagion effect can potentially spread the message to the users, making the process of adoption to be assumed by the imitators. Eventually, there will be no new clients left, hence tampering with the whole adoption process, which leads to market saturation. Moreover, there will also be a negative feedback factor related to adoption like the concerns realized from security problems, internet dynamics and IoTs. As such, internet incorporation will be dependent on the balancing aspect on negative feedback loop organization.

1.3 Literature Analysis of Next-Generation of Enterprises

1.3.1 Characteristic of the Upcoming Generation of Enterprises

In this part, we evaluate the characteristics of next-generation enterprise, which are crucial in the process of analysing the IoTs centred on ES that can mitigate problematic requirements. The characteristics include the following:

- Decentralized Decision-Making Process: The various levels and domain of organizational activities are progressively becoming more diversified compared to how they were years back. As such, hierarchical planning is applicable in effective enterprise planning for subsystem incorporation. Nonetheless, system complexities can therefore be enhanced critically based on system dynamics and scale. The centralized subsystem can potentially lead to fundamental inflexibility and time delay, which enhance the response transition quickly. In that case, decentralized and distributed architecture could be fundamental in dealing with system dynamics and complexities.
- Dynamic and Flat Structures: Timely feedback to urgent issues necessitates more decentralized and distributed organizational architecture. In that regard, the obtained information can be utilized in the process of decision-making in a timely manner. Based on the association between various networking components, one can witness various problems such as delivering data to the concerned components, particularly under central organization. The collected data is thus gathered and transferred to a centralized database before being sent to the object the moment the subsystems receive the requests from various objects. Nonetheless, the centralized structure has its own problems, especially when handling Big Data in a heterogeneity surrounding.

Based on the perspective of Big Data, the management of massive streams of data faces two challenges altering the incorporation of information systems to the upcoming generation of enterprises. These issues include the costs incurred during the process of decision-making. These costs normally increase when subsystem complexity advances. As such, this leads to resource redundancy in maintaining information locally, which is the second issue. The redundancy level wastes more resources and time for information transfer when Big Data is transferred to other decision-making structures.

Based on the heterogeneity surrounding, there is diversified and increased manufacturing resources, which have advanced the heterogeneous condition of manufacturing surroundings. These varieties exist at the facet of customized products, location distributions, regulations, cultures, suppliers, standards and optional organizations.

1.3.2   Characteristics on IoTs for Industrial Applications

The features of IoTs in industrial application include the following:

- System Dynamics: The organization of IoTs is never static; hence, it permits various system elements to be configured any moment whenever there is need to do so. Thus, this allows the integration of data over various industrial boundaries. The cornerstone industrial segment can potentially assimilate with the virtual companies, which potentially establish dynamic connections with a certain project. The structure can therefore be dismissed whenever the projects are completed, and especially when the industry should proceed to another project. Enterprises should have the capability to control the restructuring of virtual industrial alley.
- Assimilated WSNs and RFIDs Networks: These features represent one of the fundamental elements on IoTs as an information transfer protocol for data sharing and acquisition. An industrial system includes a lot of sensors to obtain machine appliances, actual-time information actuators, conveyors and features. The ancient assimilated communications are linked to peer-to-peer or point-to-point, which makes it possible to make any changes. WSN and RFID assure effective methods of supporting the decentralized and distribution of industrial resources.
- Cloud Computing: Modern industrial operation engages various activities of decision-making that necessitates intensive data and high capacities of computing. The manufacturing industries used to necessitate a lot of computing resources acting as servers in decision-making units and databases. As such, this leads to wasted investments, minimal productivity, information exchange and unbalanced utility of manufacturing elements among others. Cloud computing is advantageous since it is recommended to remedy networking issues. All the information can therefore be stored in both public and private cloud services,

which make sophisticated decision-making process to be supported by vital cloud computing techniques.

- IoTs and Humans: Associations normally take place between human, things and things, and things and humans. Various forms of these interactions are composed of various mechanisms used to support various forms of interactions. Based on the advancement of IoTs, various forms of interactions can therefore be considered under a single umbrella. In that regard, users can possibly concentrate on the various tasks without worrying about the associations. These associations are responsible for making operations and designs of manufacturing subsystems to be productive. As for the machine and human association, behaviour of humans in a virtual surrounding can be determined. Therefore, it is possible to recognize the behaviours of humans in WSN.

The developments in the IoTs have significantly led to transformative implication on the community and the environment via massive application segments, including smart agriculture, smart homes, healthcare and manufacturing. In order to completely accomplish this, a significant portion of heterogeneous IoTs appliances is networked to effectively support the actual actuation and monitoring over various domains, As such, analysts in [4] propose about 50 billion IoTs devices have to be connected by the end of 2020. Earlier on, the massive portions of data that are generally referred to as the Big Data were transferred to the cloud by the IoTs devices to enhance further analysis and processing. Nonetheless, centralized processing in the cloud cannot be relied when it comes to the analysis of massive IoTs application due to a number of reasons. One of the reasons is that applications necessitate close coupling defining the feedback and requests. The second reason is that the delays witnessed by the centralized cloud-centred deployment are not acceptable for a number of latency-sensitive applications. Third, there are extreme chances of networks failing or data being lost. Lastly, this has resulted in the transition of edge-computing remedy.

Irrespective of the fact that the present sources of literature differentiate between fog and edge computing, this research abstracts all of these cases. Moreover, novel computing paradigms are also considered as a vital concern of edge layering. The initiation of edge computing presents these major concerns based on the provision of computational capability in close proximity to information-generating devices. The smart edge network devices like the smartphones, Pi and UDOO boards provide fundamental support to local processing and information storage extensively but to a smaller segment. Nonetheless, constituent networking devices in edge computing is termed as heterogeneous as they are composed of certain architecture following a given protocol in communication. Different from the cloud where locations of data centres are fixed, IoTs and edge devices are controlled by batteries, and solar linked to an eternal power supply compared a cloud storage data centre linked to a stabilized power supply. To completely take advantage of the goodness of edge computing, it is vital to comprehend the capabilities and features of IoTs and edge devices alongside the elements of IoTs data evaluation systems. The diversification of the underlying edge and IoTs devices, formats communication mediums, formats, functional complexities and programming frameworks enhance the process of networking evaluation, which also makes it more time consuming and challenging.

Analysing the systems in an actual surrounding assures the best performance behaviour; but it is not normally easy to evaluate different frameworks in advance. Despite the availability of infrastructure, it is fundamental to perform various experimental settings since it is necessary to apply skills based on associated edge and IoTs devices that are not intuitive. Executing a lot of experiments meant to determine the correct framework necessities reconfiguration of a number of devices and transitions to necessitated parameters, which promptly become untenable appliances as a result of volume transitions required. Moreover, executing these experiments in an actual environment is significantly costly because of the maintenance and setup. Thus, the actual surrounding is considered dynamic, which means that it is difficult to reproduce much results representing various iterations leading to misinterpretation based on evaluation. These issues possibly hinder the application of actual environments used to benchmark edge computing surroundings. In order to effectively overcome this problem, a feasible alternative relates to the application of simulators. These simulators provide novel chances of enhancing the evaluation of the recommended policies and frameworks in a basic, repeatable and controlled surrounding. A simulated surrounding has to mimic the major heterogeneity and complexity of actual networks that support a lot of multiple cases affecting the deployment of IoTs.

Contrary to that, the research in [5] gives a performance analysis of the various protocols governing the application layers of IoTs structure. In that regard, these researchers provide a comparison of the Hypertext Transfer Protocol (HTTP) and the Constrained Application Protocol (CoAP) that enhances the process of formatting messages, transferring information and requesting users to evaluate various test beds. There are various deployments of evaluating various performance benchmarks for IoTs; nonetheless, this is done on the levels of platforms.

To effectively address the initial research concern, various levels and developer duties in IoTs architecture have to be considered. Figure 1.1 indicates the basic IoTs architecture (Device, Gateway and Cloud Platform), which has been diminished to three fundamental layers. In the first layer, we have the constrained controllers and devices that stand in for the IoTs [6]. Second, there are gateways, smart devices and routers that enable fog computing in each edge and can also associate as a pre-evaluated dataset from networking devices. The third layer represents the platform store, aggregate and process data retrieved from various sources, allowing enterprise application to report and analyse data to the end users. Communication and connectivity in different levels is thus not designated to a single direction. Moreover, non-functional necessities, for example security, performance and interoperability, are considered the key factors influencing the levels of performance. There are key questions that pop-up for performance engineers. These include the following:

- How can computing devices, such as disk, CPU, network and memory, be adjusted at every level?

Fig. 1.1 IoTs Architecture Abstract

- What are the implications of protocols, interoperability and security choices in reference to general networking performance?
- Do the architecture scale effectively based on the increased number of gateways and devices?

Considering the fact that IoTs stack renders more diverse, various engineers and developers are engaged in the process of implementation. Systems and embedment are concerned with the device levels, which are a bit concerned with the gateway levels. Since gateway levels continue to expand technologically, they continue to run more advanced operating systems. Moreover, application engineers and developers have continued to be a major segment of gateway levels. On this platform level, a mixture of information experts, application analysers and web developers implement the visualization and integration of information [7]. Based on the mixed interests on every level and role, it is possible to comprehend the influencing conditions in a holistic viewpoint, which has to be added to our prospective model of approach.

1.4 Challenges and Contributions of IoTs Technologies

As mentioned earlier, IoTs increases and assures the chances of linking various approaches, for instance, the approaches requiring capacity architecture for gateways and devices based on a formal model that is acknowledged in the domains of enterprise data systems. Due to the fact that there are a lot of modelling approaches, there is need to review the challenges and contributions of technologies used in the various levels.

In the device level, for example, the AutoFOCUS31 stands for the associated model-centred tools used to develop various processes in an embedded framework. This therefore includes those activities involved in evaluating modelling necessities, hardware platforms, software structure and deployment, including the generation of codes. Software architecture is created based on various software components lined to each other to enhance the interactions broken down into a lot of hierarchical components [8]. The architecture defining hardware includes different resources such as memory and processors that are connected. This also involves the platform structure applied in the runtime and execution environments like the operating architecture and Java Virtual Machines. The combination and integration of various models facilitate developers to effectively apply various synthesis and analysis techniques like model-evaluation, testing, automatic scheduling and deployment. An Eclipse Framework representing the distributed enterprise control and automation is a major segment of the Eclipse IoTs environment that represents a case for gateway level modelling. This therefore assures an open-source infrastructure defining the distributed industrial process evaluation and the control framework based on IEC 61499 protocols. To effectively model the software architecture, the 4diac incorporates an application editor, which is used to represent the functional block network that consists of a single functional block, including its various interactions based on its events. In the same manner, the separator editor is added to the model's specifications of hardware through modelling resources and devices. Through various editors and their runtime ecosystems, the 4diac permits the use of enterprise IoTs application, hence allowing interoperability, portability, scalability and configurability based on the promotion of IEC 61499 [9]. In the platform level, various associated approaches for supporting, integrating and automating engineering elements in software's lifetime are controlled using the Performance Management Work Tool (PMWT). This incorporates an automated generation of models for organization application with reference to the performance measures, modelling complex individual application behaviours and simulating the overall status of Big Data applications.

With reference to performance elements, the current remedies based on networking devices and gateway levels normally concentrate on assuring safety and functionality. Contrary to that, there are novel performance models that are used to analyse and predict the platform level behaviour. The present framework on the level of architecture assures users of the benefits this research seeks; nonetheless, we have based on enterprise application involving various necessities. For example, the workloads of enterprise applications are categorically user-initiated like the parallel user accessibilities where IoTs factors such as velocity, volume and incoming datasets are initiated by data. Moreover, parallel and massively distributed resource clusters include properties that are normally found in enterprise applications. In that regard, there is need to combine the present model techniques and the missing functionalities to enhance the performance of various models of architecture. Therefore, solvers and simulators can be implemented in deriving various metrics like throughput, response time and utility of resources. To effectively evaluate our recommended approach, we propose to model an application based on IoTs benchmarks to adapt and suit for models in different cases such as enhancing the IoTs and enhancing resource capacities. The challenges and contributions witnessed include the following.

1.4.1 Challenges

Modelling and simulating a real-life IoTs case is significantly problematic for a number of reasons. These are as follows:

- Various IoTs devices require combination of the cloud and edge systems to effectively satisfy various application requirements.
- Second, the process of modelling network graphs in various diverse forms of IoTs and edge devices is challenging.
- Third, control flows and modelling data dependencies over the edge and IoTs layer to support massing information evaluation and work flow structures are considerably non-trivial.
- Capacity evaluation over the edge computing segment depends on different configuration parameters such as data volume, upstream, downstream, data speed and bandwidth.
- Information transfer between IoTs and edge networking device is varied from cloud data centre communication that is centred on wireless or wired protocols. Connection between edge and IoTs computing layer, as evaluated in this research, is diverse. In that case, it is challenging to create an abstract without going against expressiveness.
- Mobility levels remain to be a fundamental element of IoTs devices since the sensor embedded in various physical is somewhat mobile. Due to the fact that the edge devices are considerably limited, the mobility of sensors can possibly lead to a handoff. Moreover, the information transferred to the edge devices for the purpose of process cannot be in the present range of IoTs devices. In that regard, to obtain the processed information, the edge-to-edge communication is necessary. Modelling a handoff or mobility for massive IoTs devices with various velocities is considerably problematic.
- Dynamicity of the IoTs ecosystem computes to either the removal or addition of IoTs and edge devices in most cases. This is thus caused by various factors such as network link failure or device failure. To model the scalability of IoTs devices using heterogeneous characteristics at a rapid rate is significantly problematic.
- Due to the fact that IoTs ecosystem is an upcoming development, novel applications should be created in the future. It is vital that simulators permit users to personalize their frameworks with reference to their networking necessities. Creating an overall simulator, which permits easy personalization, is considerably problematic.

Various simulators have also been proposed in past works. These include GreenCloud, and Cloud that represents the cloud environment. Nonetheless, iFog-Sim and EdgeCloudSim had been proposed for the purpose of implementation in edge computing ecosystems. However, there exists no simulator that can potentially mitigate the challenges outlined above.

1.4.2 Contributions

This research proposes the IoTSim-Edge simulators, which is meant to permit users to analyse the edge computing cases more easily since this simulator is more customizable and configurable in the ecosystem. The IoTSim-Edge is created based on the simulators that have been proposed in the past. However, the IoTSim-Edge captures the general behaviour of IoTs and the edge computing planning deployment and development. Mostly, this model deals out the challenges that have been discussed. Apart from that, this simulator can be used in analysing the present and future IoTs applications. Particularly, the model can potentially model the cases below:

- Novel IoTs application graph modelling abstract that permits practitioners to explain the information analytic operations and mapping of different infrastructure segments such as edge and IoTs.
- An abstraction supports modelling of the heterogeneous IoTs protocol with energy use profiles. This therefore permits practitioners to explain configuration of IoTs and edge devices alongside certain protocols that support the process of networking.
- An abstraction permits modelling of cellular IoTs devices. It critically captures the consequences of handoff as a result of IoTs movement network devices [10]. To effectively maintain a consistent information transfer channel, the simulator is useful for enhancing an edge communication, permitting the transfer of data to IoTs devices, that is, edge to edge.

1.5 Edge and IoTs Computing

This part provides the background analysis of the IoTs based on modelling problems. Moreover, this part also explains the overall architecture of the edge computing used to model based on the proposed simulator.

1.5.1   The IoTs Ecosystem

The IoTs are defined as 'the state-of-the-earth sensors' that are embedded in physical things and humans that are surrounding us and linked to the internet to control and monitor connected IoTs. Various IoTs application advance our day-to-day lifecycle in various vertical domains such smart healthcare services and smart homes and in disaster management. The functions of IoTs are initiated by six vital elements such as communication, sensing, identification, computation, semantics and services [11]. The IoTs devices have the capability to evaluate the ecosystem while separating ubiquitously controlling both the environmental and physical surrounding information. Thus, IoTs devices are considered more identical based on various application techniques and requirements used to implement various applications. The process of computation is distributed over various IoTs edges, devices and cloud data centre considering the desired and functional Quality of Services (QoS) and application parameters. To attain this, data is transferred from the IoTs devices to the network edge using various communication protocols. The results of computation can be utilized to initiate more decisive operations to attain the desired application processes.

Based on a simple sample of the smart homes, which control a lot of devices in urban environments, it is easier to potentially ease the lives of inhabitants. IoTs devices include the sensors added to devices such as heaters, refrigerators, cars and light bulbs, including mobile phones and gateways. The intelligent home systems utilize the private cloud information centre resources. Home appliances are linked to the gateways with light weight protocols based on the CoAP protocols enhancing the transfer of information. Mobile phones are linked to 4G whereas gateways are linked to Wi-Fi for information transfer to the private cloud [12]. In case the resident individual leaves for office, the intelligent home systems possibly switch off the heaters and light bulbs. Moreover, the systems have the capability to determine if there are things such as foods, in the refrigerator, before sending message to someone to collect them.

Modelling actual IoTs applications necessitates the linking of different actuators, sensors and edge devices on a massive scale with various operating ecosystems. This is a difficult task because of the heterogeneous features of IoTs and the edge devices requiring progressive optimization for resource allocation, migration, provision and fault tolerance in processing various applications [13]. Moreover, the process of implementation is onefold, providing a more generic framework for IoTs application in case the application necessitates level abstraction. The key problems based on modelling application, IoTs mobility, networking protocols and the consumption of energy are evaluated below.

1.6 Application Composition

Generally, the IoTs is made up of a series of tasks executed using some detected data. This can be shown using a number of techniques, although the chapter adheres to IoTs applications in a clear and (Dag) direct acyclic graph for the MELs (microelements). Every MEL is regarded to be an abstract component for each application, which stands for every assets, information and services exclusively creating the microdata, micro-service, micro-computing and micro-actuation. Therefore the modelling of every application so that they can be like the DAG for the MEL can be very difficult; hence, there is need for encapsulating a number of features altogether. Additionally, the MEL sequences have a crucial role since they are used for representing the flow of data within an abstract phase.

1.6.1 Communication Protocols

Within the IoTs surroundings, various messages and network connection protocols normally play crucial part during the communication process. Considering the genetic elements within the IoTs environment, there are various complex network linkage amidst various IoTs environment features. Centred on the range, certain limitations, device type and other protocols to be used within the IoTs network, the edged gadgets can be used for transferring information. There is need for mobile devices to leverage certain protocols unlike the static ones.

Modelling protocols using the application graph can be difficult. Moreover, some messaging processes can be accessible in order to ease the process of transferring the data from the sensors to every edged gadget and later to cloud servers. There are certain protocols that have been set for this objective. Distributing information with the aid such protocols may influence the execution processes of the systems in dissimilar ways. Therefore a single message protocol being not in a position of satisfying various requirements of the IoTs complex can also be used. For these reasons, it is necessary for every device to link up every protocol for every gadget and dissimilar layers to be used [14]. The modelling of such events using a handshaking amid these protocols can be difficult since the movement of the IoTs devices using the IoTs devices can be embedded using cars or rather smart phones that aid the various users in a flexible manner. Knowing the fact that the edged devices are normally transfixed, every IoTs device is allowed to go from one range edged device to a different resulting in a handoff. In some cases the handoff may be audible as soft or hard based on the velocity of the IoTs gadgets and the signal strength of each edge device. Therefore simulating the mobility in an adverse manner, there is need for incorporation of various features such as the IoTs devices velocity and acceleration, the pathway of the motion, topological maps, edge range intersection, etc. Taking into account all these factor, the real mobility of the simulation can be a hard role since there will be huge data number points that comprise of extensively reliant data one the relationship and characteristics [15]. Additionally, the transfer of information can be unsuccessful at any moment if the IoTs devices may be moving from a certain location to a different one basically due to poor signal.

1.7 Battery Drainage

Much IoTs drainage is driven by the battery, which is normally a restriction and hence should be recharged at various intervals. Hence it is crucial that the devices should be in a position of holding off the battery for an increased time mostly when the applications cannot be very easy recharging, for example, the sensors within rivers or places of havoc. Holding the battery off for an increased period of time will highly save much time, and this is crucial almost for every application. Therefore transferring information at dissimilar intervals and utilizing various communications protocol.

1.7.1 Architectural Computing for IoTs Edged Devices

As shown in Fig. 1.1, the architecture of the IoTs edge has been shown in detail. Moreover the IoTs technology comprises of a number of components such as the sensing nodes and also actuator nodes. The sensing nodes are responsible for the collection of data within the sensors and transmit the data whereby it will be required to be processed and stored. The actuators will therefore be centred on the data analysis [16]. Thus the subsequent layer will be for the edged infrastructure that comprises of different forms of edged devices, for example, the arduino and raspberry Pi. Such gadgets may be easily accessed openly with the aid of various forms of containerization and virtualization techniques. These mechanisms provide sufficient infrastructure that is used for the deployment of various raw data created through various sensor nodes. For many instances, if the edge can be in a position of creating more data, then there is no need of transmitting the information via the cloud for processing. Lastly, its outcome is later transmitted to its actuator so that a certain action can be executed. Therefore the services and the application layers comprise of dissimilar services which can be accessed to various clients. Later the applications can be acquired via a subscription channel. Some of the examples for these services include smart city and smart home. Every MEL is normally distributed within an edge or rather the cloud information centre. Additionally, the program helps in managing every QoS necessity within the prospective load and any error handling. Moreover, it delivers services, for example management of resources, device management and storage management.

Service issued by such a layer normally ascertains that the QoS needs have been successfully met. The present application distribution and scaling methods created for the distributed environmental algorithms, for example, the cloud or grid, cannot be effective when employing the new IoTs environment. The main reason is there are extensive features and characteristics used for the smart gadgets aside the mobility elements and the latest applications technology which is made of a limited reliance and hence needs process distribution. Based on the type of application, there has been collaboration amidst the IoTs, cloud and the edge that are required for attaining a prospective QoS desires. Therefore, creating new applications deployment alongside its scaling methods is very crucial. Moreover, it is important to analyse and test such methods prior implementing the actual deployment. Nevertheless, analysing these kinds of techniques require a real environment incorporating a set of conditions at every moment which is expensive and time consuming. Additionally, based on the alignment of ownership of various gadgets, assessment needs a number of mechanisms that may make something become much difficult. Hence, various simulation frameworks, for example, the IoTSim simulation that aids the distribution of these applications that assess the execution processes various methods, certain events within certain conditions may be necessitated [17]. Additionally, assessing these methods in certain circumstances may be executed at low cost within a simulation environment.

1.7.2   The IoTs Sim Edge Architecture

Based on the architecture for the prospected simulator, it shows a series of layers. A concise analysis of its distinctive components has been highlighted in this section. IoTs Sim edge is structured within clouds simulation equipment. The cloudsim delivers an underlying system that is used for assisting some basic communication processes within various subscribed components that use the event management mechanisms. Hence the primary components used for the cloudsim have been extended so that they can stand in for the edge infrastructure within a line comprising edged characteristics and features. The IoTs asset layers comprise dissimilar forms of IoTs gadgets in which they have their own elements and characters alongside execution processes used for actuation and sensing [18].

1.7.3   Implementation and Design

When the IoTSim edge is enacted, the missing gap amidst the cloudSim has to be extended alongside vast numbers of classes so that the model can be shown in an edged and IoTs environment. A certain entity that increases the SimEntity class may seamlessly transmit and acquire various events for its partners via the event management engine. Moreover, this research exhibits a basic trait for the simulators, a cycle that comprises of various actions and procedures. Therefore, to model the edge infrastructure, there was need for the designation of vast number of classes within the network. Major classes include the edgebroker, edgedevice, edgedatacentre, microelement, edgedatacentrecharacteristics and the edgelet. For instance, the edgedatacenter's role is used for creating a link between the IoTs devices and edge centred on some provided IoTs protocols, and the process is accompanied by asset edge provisioning scheduling rules and assessment of edge processes. Moreover, the system is also responsible in the creation of the edgedevice, submission of the edgelet, establishing of a network amidst the IoTs infrastructure, assessing network accessibility, etc. These particular characteristics for the edgedatacentre are always fed using the edgedatacenter characteristics class [19].

For this reason, the edgebroker's role is acting in place of the clients based on establishing a connection from the IoTs devices and the edge devices, submission of the edge and IoTs requests, and acquiring various outcomes. This class outlines a model for the edged gadgets specifically when acquiring and processing information from the IoTs in which the data processing is executed following a specific edgelet rules and policies. On the other hand, the MEL class is used for modelling the abstract processes that later performs within the IoTs data for each edged side, or every cloud information centre. Considering the present implementation for each MEL, only the running of the edged devices will be prioritized. Hence the edgelet model class is required to be performed within the MEL. In some situations the edge device could consist a battery such as a mobile phone, hence going around, thus, the mobility and battery class will be designated to activate the edged device so that it can extract its constituent characteristics. Therefore, the Moving Policy class prescribes the conditions and characters for each edge device [20]. At the same IoTs infrastructure can be modelled, vast number of new classes have been designed.

Regarding speed rate such as Wi-Fi, the Network Protocol is a recognized communication system where it can transfer network packets at a velocity of 200 Mbps, and the 4G LTE will be at a speed of 150 Mbps. Hence the frequency for the edgelet in different terms, the delay period that will be required to transmit the edgelet to the datacentre that is normally accessed from the network policy class. On the other hand, the class models for every feature within the IoTs protocol in relation to the battery and QoS consumption rate has been mentioned [21]. Since each IoTs process comprises of dissimilar processing methods, everything is structured in a manner that they may have dissimilar energy consumption rate. Much more detailed explanation based on the classes has been given in the following sections.

1. EdgeDataCenter Class

This class is responsible for controlling various core edge infrastructures. At every stage counter-intercept various incoming calls and processes and execute dissimilar protocols centred on the payload of a particular event, for example asset delivering and handing over various edglet requirements to its desired MELs. When the battery of every edged device is discharged, the edgedatacenter will detach automatically and send various unserved requests to its subsequent edged devices.

2. EdgeBroker class

In this class, there are some clients' proxy, whereby various requests are created focusing on the prescribed necessities. This class comprises of a range of roles for executing, for example, the submission of edge and provisioning of the MEL requests to every datacentre, getting the IoTs devices to create and transmit data to its prospective edged devices and acquiring its final processing outcome. The class aids various power aware models for every IoTs gadgets. Since the edge broker consistently assesses the consumption of every battery for the IoTs gadgets, immediately there will be a disconnection from the nearby IoTs devices.

3. EdgeDevice class

The class is assumed to be the same as actual edge devices. Its main role is hosting a number of MEL and simplifying the protocols of the CPU sharing appliances through a specified CPU sharing rule. Moreover, they are further linked to have a certain number for each IoTs device that transmits its data for processing purposes. Focusing on the present version of the simulator, a set of four classes has been used to extend the IoTs devices class which include lightsensor, voicesensors, carsensors and temperature sensors. Therefore, an improvised form of IoTs device may easily extend the IoTs device class and hence enact a new set of features. Within an actual IoTs environment, each IoTs gadget harbours an IoTs and network protocol; hence, the class comprises of same features when utilizing the networkprotocol and IoTProtocol classes as analysed in subsequent phases in this chapter.

4. MEL class

The role of this class is representing a single component using the IoTs application graph. It basically represents the core processing necessities used by the application components. Centred on the application requirements, setedgeoperation techniques may be configured. Hence the reliance amidst certain components may be shown by using an uplink and downlink that can be easily structured in order to portray some possible difficult applications [22]. This can help in configuring which stands in for any difficult IoTs applications centred on the clients necessities.

5. The EdgeLet class

The class models every IoTs created data and its subsequent MEL processing information. At any moment that the IoTs gadget created a unique link using its respective edged gadgets, there will be an immediate response that will generate IoTs sensed data that will be required in the form of the EdgeLet which comprises of a payload. The role of a payload is encapsulating the correct data so that it may direct the MEL at the processing podium, for example, the size of each data set created, the route followed by the MEL ID, and the IoTs device ID [23]. When using the size of the EdgeLet data size, the delay of every network will be needed to transmit the EdgeLet to its required MEL and hence can be calculated considering the network transmission frequency.

6. Mobility class

The class is made up of a model of IoTs and alongside various edge devices. Its mobility has a crucial role within the IoTs edge system in which upholding actual time consistency is crucial so that the performance processes can be executed. Its prospective attributes may comprise of velocity, location, range and the time interval. Every attribute taking off time comprises of some distinctive values so that they may represent the vertical and horizontal direction for every edged IoTs devices. Moreover, this chapter portrays some possible examples of the way various attributes have been utilized using the edge datacentre so that the location can be reviewed by the edge and IoTs devices. On the other hand, algorithm 1 portrays pseudo-code used for simple tracking methods enacted within the EdgeDataCenter class. Various clients may easily increase every class so that they can implement their mobility models.

7. The Battery class

This class is used for modelling the battery features for the portable IoTs and the edge devices. Through using the transmission rate modelling, the time used for transmitting every data may be achieved considering the size of each EdgeLet.

8. Policies classes

The main functions of these classes are modelling the policies in three distinct groups, which include: battery consumption, device movements and network transmission. Therefore, the device movement rules may instruct edge datacenters so that there can be tracking of the location and the movement of both the edge and IoTs devices. The consumption of the battery rules normally calculates and tracks the subsequent power capabilities for each edged and IoTs devices. The transmission of the network policy calculates the duration of time taken so that the data can be distributed from one IoTs device to another edged device [24]. These kinds of policies may be lengthened so that a new IoTs edged solution and strategy can be derived. Both the classes may be extended so that various client policies can be extended and implemented.

9. User Interface Class

This class normally provides the required techniques that can aid effective configuration for each IoTs applications not knowing other data based on the simulators. Moreover, it permits the user explaining every parameter while using associative interfaces that can be converted to a required configuration file. The content below shows the data in the user interface.

```
"ioTDeviceEntities" : [
{
"mobilityEntity" : {
"movable" : false,
"location" : {
"x" : 0.0,
```

"y" : 0.0,
"z" : 0.0
}
},
"assignmentId" : 1,
"ioTClassName" : "org. edge. core. iot. TemperatureSensor",
"iotType" : "environmental",
"name" : "temperature",
"data _ frequency" : 1.0,
"dataGenerationTime" : 1.0,
"complexityOfDataPackage" : 1,
"networkModelEntity" : {
"networkType" : "wifi",
"communicationProtocol" : "xmpp"
},
" max _ battery _ capacity" : 100.0,
"battery _ drainage _ rate" : 1.0,
"processingAbility" : 1.0,
"numberofEntity" : 5
}]

1.7.4   Computation and Event Processing

Various simulation processes occur on time for initializing the proper IoTs edge structure that has been extracted for a certain file configuration. Moreover this research exhibits the configuration test for every file within the IkT gadgets. After the creation of the MELs and the IoTs gadgets, every edge broker will request data from its centre so that effective connections can be established amidst the IoTs devices and hence its respective MELs.

At any moment that the edge broker gets the new established connection (ACK), there will be an immediate notification from the IoTs devices based on the connection. Moreover, the IoTs devices can commence creating data as edgelets, distributing the edgelets to its final MELs and reducing the battery power based on the rate of drainage. Hence the IoTs devices can maintain the execution processes (Fig. 1.2).

1.8   Case Studies: Sim Assessment for the IoT

In order to assess the usefulness of the IoTs sim edge, three scholarly studies have been carried out. Explanations of each study has been identified in the below case.

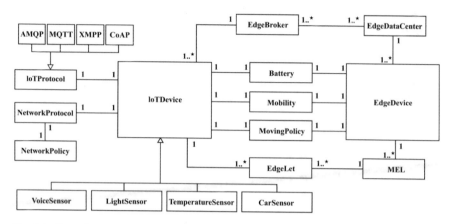

Fig. 1.2 Class diagrams of the IoTSim-Edge model

1.8.1   Case One: Healthcare Sector

Considering various works that involve activity information has been segmented into various micro-applications, there were also some realizations about the similar circumstances. So that the raw information utilizing the phase counts calculation, every micro-operation is represented to be as the MEL which can be undertaken by the unforeseen IoTs gadgets. There is a basic MEL graph portraying the one necessary for deploying the edge within computing vicinity. For finding out the pareto-customized deployment answers within a deployed IoTs infrastructure, it is required to evaluating dissimilar deployment strategies for the matter to be effective. Due to the fact that there has been limited reliance within the MEL, it is crucial considering such a factor during the deployment process. In order to enact the main simulator, two-edged gadgets can be embedded within the IoTs devices. For these reasons, the devices create information centred on some elaborated data frequency.

Scientific evidence is evidence that serves to support the rate at which batteries drain that the rate of transfer is always greater compared to its algorithm. For the fact the edged devices powered by a battery, the process that happens within the device edge, then the rate at which every battery drains is handled in this chapter, although executing much more performance processes close to the IoTs gadgets can augment the performance time that is crucial in many instances. Additionally, various outcomes have shown and analysed overall performance duration in dissimilar scenarios. Centred on actual scenarios, the MLs have been allocated dissimilar MELs. Moreover, this research evaluates a quick rational about battery hours that comprise of edged processes which demand a shrinking element. Hence various shrinking features have occurred within the edged gadget.

Hence the result performance rate for every edged devices was 90% that was embedded which results to saving much more battery power unlike conveying raw information to the rest of the devices. Therefore the outcome for every processing time is 90% for each execution of the device which maximizes the utilization of the battery by 266% unlike executing 10% operation. For the fact that the edge E1 comprises of a minimal processing and harbouring power, every process may not be explained in this section. Dissimilar forms of evaluation may be executed by using various scenarios. Clients may additionally suggest dissimilar computations so that an effective deployment strategy can be found to focus on quite a number of objectives found.

1.8.2   Case Two: Smart Building

Previously, smart building systems automatically caused the heating, lighting, air-quality, air-conditioning, etc., which have gained much focus. Various forms of sensors that transmit at certain intervals and specific sites have the capacity of sending information to various linked devices that may process and analyse various data sets. Therefore edged gadgets conduct common processes and transmit information to the cloud and when this happens there is increased storage or complex analysis necessitated. Various IoTs devices transmit its data to edged devices complying with a common communication process. Various elements such as latency imminently rely on the rate of the data and size of the packet [25].

1.8.3   Case Three: Capacity Planning Units for Roadside Self Driving Cars

Some of the incoming advancement whereby every vehicle can be embedded sensors that transmit data to be RSU controllers which consider the consistency of the traffic and maintain the safety of the road. Moreover, the RSUs may provide an excellent platform for the edged devices to transmit the data acquired from a specific vehicle and may create a runtime decision. The car ranges alongside the RSUs are restricted, and hence every vehicle is directed at a certain speed to a certain point of the road. Therefore the area covered by the RSU solely relies on the protocol transmission. Probably, the car's connection diminishes alongside the linked RSU at a point of time. At some moment the handoff can be difficult following the previously mentioned RSU so that various decisions can be made, while an RSU-RSU information delivery will also be created.

Based on the point of focus of the car, an RSU may convey a message to a different RSU that can be later directed to the final car. Because the Gap and the processor capability of the RSU are always restricted, evaluating the count of vehicle data to be processed is possible by making sure that there is no data loss. Additionally, it is essential to assure that the requirements for the QkS such as the application time response are available. The scenario was effectively structured using a simulator. At first there is an establishment of a brand new link for the RSU referred to as RSU1, which commences transferring information. On the other hand, at time t3, a certain point is reached whereby the range of both RSUs is attained. Relying on its mobility, a connection can be created by RSU2 but is still transmitted to RSU1. Regarding its motion, the connection can be made, hence data will be sent to RSU2. Moreover, RSU1 finds out that every car has its range and can be centred on the data being transferred as exhibited by time t4. Such an event stipulated that these mobility features are cooperative when accessing days pertaining every simulator.

On the other hand, precise configuration of each simulation has been considered in this research specifically. These edged devices are unique except the location which includes E1: 0, 0, 0 and E2: 50, 0, 0. Based on this fact, the outcome highlighted can precisely prove the idea [26]. Because the edged devices number are always constant and the processing manner is timely, various requests may congest prior processing, resulting in a greater performance processing time. Furthermore, when a certain vehicle goes further away from the RSU, therefore the RSU need to transmit the data to the other RSU whereby the data is later transmitted to the other cars. Because such comes among the processing time, hence performance period will be great. Therefore the greater the number of various requests for a certain care, then the edge can take up great energy as assessed by these simulation that portrays the averagely energy utilization rate for the edged devices as they increase from one number to the other within the IoTs network.

1.9 Related Sources of Literature

If there is much interest in the computing edge of the IoTs, vast simulation equipment would have been created for previous years. Various tools can be extended considering the present network and the loud simulator, although there exist some gap within the simulator and actual modelling edge and the vicinity of the IoTs. This phase analyses the present simulation equipment used within the network, IoTs environment, cloud and the way they cannot model the present IoTs edged environment. Moreover, it is shown that the simulation structure is in a position of meeting the expectation of the accessible limitations using holistic methods.

1.9.1 Network Simulators

Some tools used for the simulation process have been recommended and hence preferred to be used within the computer networking system for the last years. Considering all this, this section analyses some common recognized network simulation equipment. These tools include the C++ and the OMNeT++ focusing on the discretion of various events of the environment alongside the communication network. The system aids a parallel simulation network and a real-life implementation protocol for the simulation models. Additionally, various network processes can be supported although cannot harbour the edge communication protocols [29]. According to scientific analysis the OMNeT++ system was created so that there can be consistent simulation body area network and some accompanying lower powered gadgets. Furthermore, the simulation can be endured within the network with an extensive number being dynamically driven. By increasing the energy support for every model different protocols for the communication process can be used. An example of a source-driven simulation includes the TOSSIM, which is used within the WSN (wireless sensor network). Therefore, the energy usage may not use any mobility modelling and its consumption. In order to aid energy consumption rate within the WSN, a different simulator has been suggested, referred to as the powerTOSSIM, which is achieved by extending the TOSSIM. Therefore, the powerTOSSIM takes into account various nonlinear characters for the battery simulation model.

Furthermore, the mobility elements have not been mentioned. Whereas the NS-3 will a different C++ model for a simulation whereby it is made up of the python interface. Such equipment is used for the simulation that normally distributes visual aid to the neighbouring environment. The NS-3 will not be an effective IoTs simulation within an edged level as they do not provide support for the application and scheduling of various features. For the cloud simulation, majority of them have been recommended to be used as the best cloud computing tool which provides excellent outcomes [27]. Amidst them, the cloud sim has been regarded to be prominent simulator so far within the research sector. Various clouds' sim that are event driven and that aid the system and behavioural modelling within the cloud environment have been mentioned. Nevertheless, the IoTsim edge has been created to aid extensive scale simulations whereby there can be support for various communication systems and also physical modelling. Such simulators comprise of customized international hypervisor that can aid cloud brokering rules. Generally it may comprise of the Amazon public clouds scenarios and aid the MPI. Additionally the network cloudsim can be a different simulation that permits such models within the network that comprises of cloud information centres [28].

In these cases, such simulators may not be in a position of aiding the IoTs and the edge simulators. A different cloud simulator includes the green cloud that can be increased using the NS2 simulation equipment. On the other hand, greencloud is made up of packet-level equipment which is used for establishing the amount of power usage for the components within the data centre. The main functions of the simulator are computing the energy used in order to ascertain that there is energy awareness placement. Moreover, there is no support of the edge and the IoTs simulation. The DCSim is a different cloud simulation that entirely reflects on the IaaS cloud vicinity simulation. The main role is aiding the simulation and modelling of various data centres, the VM and the host using restricted application number and management of asset policies. Conversely, the IoTs simulation and the edged environment cannot be aided using this type of simulator. Much simulation equipment has been recommended for the simulation within an edged environment. The SimIoT is known to be simulation equipment that effectively models the communication amidst the cloud data centres and the IoTs devices, although they do not take into account various edged device within the simulation [30].

Generally, not all simulators support the edge information transfer protocols and the battery power. Moreover, most of these simulators are not capable of defining the application composition in the IoTs environment. These characteristics are vital for use in any IoTs application. The composite simulation ecosystem of the Edge and IoTs is vital to researchers and enterprises since it helps them to gain actual control of edge processing.

1.10   Conclusion and Future Work

In conclusion, this chapter pursues and recommends a model-centred approach used to evaluate and predict the performances of IoTs systems and architectures. The IoTSim-Edge model shall potentially allow engineers and developers to evaluate various designing choices in systems, in advance. Moreover, the model will effectively locate possible bottlenecks, size the resources and predict any potential feedbacks to visualized results. The future prospects are bases of works that integrate and combine various model approaches used to embed systems with Big Data frameworks and enterprise data systems. In that regard, there is need to formulate a prototype that will be used to intergrade modelling ecosystems. The present manufacturing ecosystems have been evaluated to effectively analyse the vital necessities of ESs based on modern organizations.

References

1. Anandakumar, H., & Umamaheswari, K. (March 2017). Supervised machine learning techniques in cognitive radio networks during cooperative spectrum handovers. *Cluster Computing, 20*(2), 1505–1515.
2. Abbas, N., Nasser, Y., & Ahmad, K. (2015). Recent advances on artificial intelligence and learning techniques in cognitive radio networks. *EURASIP Journal on Wireless Communications and Networking, 2015*(1), Available: https://doi.org/10.1186/s13638-015-0381-7.
3. Belkin, M., & Niyogi, P. (2004). Semi-supervised learning on Riemannian manifolds. *Machine Learning, 56*(1–3), 209–239. Available: https://doi.org/10.1023/b:mach.0000033120.25363 .1e.
4. Benhamou, E. (2019). Similarities between policy gradient methods (PGM) in reinforcement learning (RL) and supervised learning (SL). *SSRN Electronic Journal.* Available: https://doi.org/10.2139/ssrn.3391216.
5. Cecin, F., Barbosa J., & Geyer, C. (2006). A communication optimization for conservative interactive simulators. *IEEE Communications Letters, 10*(9), 686–688, Available: https://doi.org/10.1109/lcomm.2006.1714546.
6. Chang, X., Mi, X., & Muppala, J. (2013). Performance evaluation of artificial intelligence algorithms for virtual network embedding. *Engineering Applications of Artificial Intelligence, 26*(10), 2540–2550, Available: https://doi.org/10.1016/j.engappai.2013.07.007.

7. Anandakumar, H., & Umamaheswari, K. (October 2018). A bio-inspired swarm intelligence technique for social aware cognitive radio handovers. *Computers & Electrical Engineering, 71*, 925–937. https://doi.org/10.1016/j.compeleceng.2017.09.016.
8. Farhat, S. (2018). Resource sharing in 5G multi-operator wireless network. *International Journal of Digital Information and Wireless Communications, 8*(3), 156–161, Available: https://doi.org/10.17781/p002431.
9. Fu, Y., Wang, S., Wang, C., Hong, X., & McLaughlin, S. (2018). Artificial intelligence to manage network traffic of 5G wireless networks. *IEEE Network, 32*(6), 58–64, Available: https://doi.org/10.1109/mnet.2018.1800115.
10. Gueguen, C., Ezzaouia, M., & Yassin, M. (2016). Inter-cellular scheduler for 5G wireless networks. *Physical Communication, 18*, 113–212. Available: https://doi.org/10.1016/j.phycom.2015.10.005.
11. Gueguen, C., Ezzaouia, M., & Yassin, M. (2016). Inter-cellular scheduler for 5G wireless networks, *Physical Communication, 18*, 113–124, Available: https://doi.org/10.1016/j.phycom.2015.10.005.
12. Haldorai, & Kandaswamy, U. (2019). Supervised machine learning techniques in intelligent network handovers. *EAI/Springer Innovations in Communication and Computing*, 135–154. https://doi.org/10.1007/978-3-030-15416-5_7.
13. Huang, R. P., Zhou, & Zhang, L. (2014). A LDA-based approach for semi-supervised document clustering. *International Journal of Machine Learning and Computing*, 313–318. Available: https://doi.org/10.7763/ijmlc.2014.v4.430.
14. Iosifidis, A. (2015). Extreme learning machine based supervised subspace learning. *Neurocomputing, 167*, 158–164. Available: https://doi.org/10.1016/j.neucom.2015.04.083.
15. Kansal, P., & Shankhwar, A. (2017). FBMC vs OFDM waveform contenders for 5G wireless communication system. *Wireless Engineering and Technology, 08*(04), 59–70. Available: https://doi.org/10.4236/wet.2017.84005.
16. Kaur, K. S., Kumar, & Baliyan, A. (2018). 5G: A new era of wireless communication. *International Journal of Information Technology*. Available: https://doi.org/10.1007/s41870-018-0197-x.
17. Ke, C., & Shihai, W. (2011). Semi-supervised learning via regularized boosting working on multiple semi-supervised assumptions. *IEEE Transactions on Pattern Analysis and Machine Intelligence, 33*(1), 129–143, Available: https://doi.org/10.1109/tpami.2010.92.
18. Khanale, P., & Chitnis, S. (2011). Handwritten Devanagari character recognition using artificial neural network. *Journal of Artificial Intelligence, 4*(1), 55–62. Available: https://doi.org/10.3923/jai.2011.55.62.
19. Khanale, P. (2010). Recognition of Marathi numerals using artificial neural network. *Journal of Artificial Intelligence, 3*(3), 135–140. Available: https://doi.org/10.3923/jai.2010.135.140.
20. Ma, J. Y., Wen, & Yang, L. (2018). Lagrangian supervised and semi-supervised extreme learning machine. *Applied Intelligence, 49*(2), 303–318, Available: https://doi.org/10.1007/s10489-018-1273-4.
21. Manasa, H. R., & Pramila, S. (2015). Architecture and technology components for 5G mobile and wireless communication. *International Journal of Engineering Research and, 4*(06), Available: https://doi.org/10.17577/ijertv4is061112.
22. Mythili, K., & Anandakumar, H. (December 2013). Trust management approach for secure and privacy data access in cloud computing. In *2013 International Conference on Green Computing, Communication and Conservation of Energy (ICGCE)*, doi:https://doi.org/10.1109/icgce.2013.6823567.
23. Nastase, V., & Strube, M. (2013). Transforming Wikipedia into a large scale multilingual concept network. *Artificial Intelligence, 194*, 62–85. Available: https://doi.org/10.1016/j.artint.2012.06.008.
24. Nishii, R. (2007). Supervised image classification based on statistical machine learning. *SPIE Newsroom*, Available: https://doi.org/10.1117/2.1200612.0449.

25. Sądel, B., & Śnieżyński, B. (2017). Online supervised learning approach for machine scheduling. *Schedae Informaticae, 12016*, Available: https://doi.org/10.4467/20838476si.16.013.6194.
26. Shang, Z. (2005). Application of artificial intelligence CFD based on neural network in vapor–water two-phase flow. *Engineering Applications of Artificial Intelligence, 18*(6), 663–671. Available: https://doi.org/10.1016/j.engappai.2005.01.007.
27. Singh, P. (2016). Importance of wireless sensor network and artificial intelligence for safety prerequisite in mines. *International Journal of Engineering and Computer Science*. Available: https://doi.org/10.18535/ijecs/v5i11.35.
28. Singh, P. (2016). Safety of underground coal mine using artificial intelligence and wireless sensor network. *International Journal of Engineering and Computer Science*. Available: https://doi.org/10.18535/ijecs/v5i11.38.
29. Vishnoi, R., Gaur, S., & Verma, A. (2018). Comparison of various waveform contenders of 5G wireless communication based on OFDM. *International Journal of Trend in Scientific Research and Development, 2*(4), 2468–2473, Available: https://doi.org/10.31142/ijtsrd15637.
30. Zhao, H. (2018). Semi-supervised extreme learning machine using L1-graph. *International Journal of Performability Engineering*. Available: https://doi.org/10.23940/ijpe.18.04. p2.603610.

Chapter 2
Organization Internet of Things (IoTs): Supervised, Unsupervised, and Reinforcement Learning

A. Haldorai ⓘ**, A. Ramu** ⓘ**, and M. Suriya** ⓘ

2.1 Introduction

Internet of Things (IoTs) links up various data sensing appliances via the Internet, which is meant to realize smart management and identification. Different intelligent sensors are a vital building block used in establishing Internet of Things (IoTs)-based business process applications. In this case, the business process management (BPM) framework is not only relevant in enhancing the proficiency of collaboration in cross-sensing organization procedures, but it is also relevant in assisting to invoke effective management competencies before an emergency snowballs into disaster, i.e., traffic jams, fire hazards, or networking failures. Due to these diversities in sensing and the variations of their different functions, the direct models of the IoTs-centered organization processes and applications are significantly difficult. Developers require using more coding practices, which bring out the aspect of ignoring organizational logic orchestration. Perhaps, the fundamentals of the IoTs applications before organizational procedure automations represent the creation of organizational procedure models. These models are computed based on all forms of organizational procedure obligations that signify novel functional units that have been implemented by various services. The BPM frameworks shall benefit from representing modeling when fundamental sensor devices can assume the position of sensor resources for personal organizational procedures.

A. Haldorai (✉)
Sri Eshwar College of Engineering, Coimbatore, Tamil Nadu, India

A. Ramu
Presidency University, Bangalore, India

M. Suriya
KPR Institute of Engineering and Technology, Coimbatore, Tamil Nadu, India

© Springer Nature Switzerland AG 2020

27

A. Haldorai et al. (eds.), *Business Intelligence for Enterprise Internet of Things*,
EAI/Springer Innovations in Communication and Computing,
https://doi.org/10.1007/978-3-030-44407-5_2

The Internet of Things (IoTs) is confirmed as the distributed and interlinked network of embedded frameworks that communicate via wireless or wired communication advancements [1]. As such, this has been confirmed to be the network of physical things and objects that have been empowered with limited storage, communication, and computation capacities, which include embedded electronics like actuators and sensors, network connectivity, and software that are vital for objects used to gather process and transfer information. The IoTs represent the objects realized from various life scenarios that include intelligent devices kept in buildings. These devices include smart adapters, smart bulbs, smart refrigerators, smart meters, and smart detectors. Other more sophisticated devices include the radio-frequency identification devices (RFID), accelerometers, heartbeat detectors, and sensors located in parking lots, among others. There exist a lot of services and applications provided by the Internet of Things (IoTs) that range from novel infrastructure to military, agriculture, personal healthcare, and home appliances. Moreover, the domains dealt by the Internet of Things (IoTs) services include medical, energy, retail, building management, manufacturing, and transportation, among others.

A massive scale of the Internet of Things (IoTs) network utilizes novel approaches like the management of various devices such as the absolute scale of information, communication, storage, security, privacy, and computation. Up to this moment, a significant research analysis has been conducted concerning the various elements of IoTs such as communication, architecture, applications, protocols, privacy, and security. Nonetheless, the foundation of IoTs commercialization and its technological advancements are centered on privacy and security guarantees, which include user satisfaction. The idea of IoTs using the enhanced technologies like the software-defined networks (SDNs), fog computing (FC), and cloud computing (CC) means that landscaping the threats for attackers has advanced. The generation of information based on IoTs devices is significant, and therefore, the ancient information collection processes and storage techniques are not operated under this scale. Moreover, the absolute information set can be utilized based on behavioral controls, patterns, assessments, and prediction techniques. Moreover, the data heterogeneity produced by the Internet of Things (IoTs) formulates more fronts proposing the present information processing techniques [2]. As such, to effectively harness the worthiness of information retrieved through the IoTs, there is a need for the implementation of novel techniques. Resultantly, it is considerable to conclude that machine learning (ML) is a vital computation paradigm used to provide embedded smart devices based on IoTs.

The idea of ML is fundamental for intelligent devices and machines to inject vital competency from humans and device-generated data. Moreover, this idea can be referred to as the competency of intelligent devices to automate and adapt to various behaviors or situations based on knowledge, and this is viewed as the fundamental element of implementing IoTs solutions. Thus, the ML technologies have been utilized in obligations like regression, classification, and estimation of density. Different applications like fraud detection, computerized visions, malware detection, and speech recognition make use of ML technologies and algorithms. Similarly, these technologies are used to leverage the idea of IoTs in availing smart services to users.

2.2 Radio-Frequency Identification Devices of Internet of Things (IoTs) Features

2.2.1 Internet of Things (IoTs) Network Features

In this section, we shall evaluate the novel features of Internet of Things (IoTs) networks. These features include the following:

Heterogeneity: The Internet of Things (IoTs) networks include a lot of devices that are equipped with various capacities, information transfer rules, and features of communication that relate to one another. To put this into simple language, the devices can utilize various forms of information transfer techniques, including the various communication paradigms like Ethernet and cellular paradigms. Moreover, the variable constraints concerning the hardware resources are also some of the techniques considered.

Large-scale deployment: It is considered that thousands of devices linked to one another via the Internet shall surpass the capacities of the present Internet conditions. The large-scale deployment of Internet of Things (IoTs) also proposes significant issues. These issues include the designing of networks and their storage architectures for intelligent devices, information transfer protocols, and effective data sets. The effective data sets communication rules, proactive protection, and identification of Internet of Things (IoTs) are fundamental for securing against malicious attacks, technological standardization, and management of application interfaces.

Inter-connection: The Internet of Things (IoTs) is purposed to be linked with international data and communication infrastructure, which can be retrieved from any location and at any moment. This form of connection is dependent on various applications and services that are produced by IoTs service providers. In some moments, the connection can be termed as local while in other cases it is considered to be international. The local cases include the instances of interlinked motor technologies and swarm sensing [3]. On the other hand, the international cases will include the intelligent home accessibility via the mobility infrastructure and crucial management of infrastructure.

Information transfer in close proximities: One of the most outstanding features of the IoTs is information transfer in close proximities that has to eliminate the centralized standards such as base stations. The device-to-device (D2D) information transfer influences the features of on-point information transfer, for instance, the dedicated short-range communication (DSRC) and many other technological advancements that can be rated the same. The architectural segment of the ancient

Internet is considered to be inclined more on network-centric information transfer as compared to how communication presently has been delinked to complement the services of the IoTs spectrum.

The ultra-reliable and low-latency communication (URLLC): The URLLC is considered as another feature of the IoTs networks necessitated in novel actual-time applications and services such as organizational procedural automations, smart traffic, remote surgery, and smart transportation frameworks. The vital performance limitations include reliability and delay aspects.

Inexpensive and less-energy information transfer: More connectivity linking the IoTs devices necessitates ultra-less-energy and inexpensive remedies that facilitate smooth operation of networks in the modern age. Individual-organization and individual-healing features are necessary for contemporary and urgent IoTs information transfer, which include disaster and emergency conditions [4]. In these cases, the reliance of networking infrastructure is never considered an option, which means that individual organization of networks has to be considered. The vibrant transitions in networking are composed of a significant number of devices, which required management in the most effective manner. All these devices have the capability to act vibrantly. For example, the wakeup or sleep duration of devices shall be dependent on various applications, including the time when these devices utilize the Internet.

All these features have to be included in IoT networks. One of the vital features among them considers the safety of networks. Among these features, however, safety is a vital element that enhances the smooth operation of IoTs networks. The safety of networks is considered to be significant not only for the devices but also for the consumers. This is due to the fact that IoTs devices linked to the Internet can potentially be tampered as individual information has been shared with different devices. Moreover, the security and privacy of these devices is a fundamental factor that has to be considered in networking.

Smart networking is also an intriguing feature of Internet of Things (IoTs), which enhances informed and timely decisions that have to be executed in organizations. The information generated by IoTs devices has to be created in a manner that facilitates the performance of actions to effectively make decisions enabling the processing of information [5].

The intelligent urban environments utilize modern data and communication technologies to analyze and integrate the information gathered from the vital frameworks that navigate through the urban environments. At the moment, the intelligent cities are capable of executing smart responses to different cases such as weather forecasting, traffic controls, and organizational and economic activities. An intelligent urban environment with its traffic-routing system entails a massive number of cameras that are meant to monitor the road networks and the intelligent algorithms that navigate across the cloud networks proposing the optimum routes used for individuals. Moreover, the intelligent motor navigation framework permits individuals to change and set up destinations through the inbuilt audio appliances. The two frameworks in pairs are used to provide actual-time interactive routing systems. Nonetheless, the individuals' voice commands can be translated into motor edges and their sides before they are transferred into cloud systems where intelligent routing frameworks operate. The most vital route can be translated to the voices that guide individuals to their preferred destinations. The applications that have been mentioned above are utilized in different computing resources such as edge, IoT devices, and cloud services, which include modern language techniques that enhance developments of the various ML-centered IoTs applications [6]. These advancements are challenging for modern language models and IoTs frameworks. To deal with the available gap, this research has orchestrated the enhancement lifecycle of the modern language-centered IoTs applications. In the subsection outlined below, we have included an analysis of the enhancement lifecycle alongside the detailed taxonomy, which surveys the techniques needed in the enhancement of applications.

2.2.2   Privacy Issue of Internet of Things (IoTs) Deployment

Privacy and security issues are the vital factors that enhance commercial realizations of Internet of Things (IoTs) applications and services. The present Internet setting is an attractive segment for privacy concerns that are relevant for a few tasks, which includes the corporate levels and well-linked privacy breaches that have a significant effect to various organizations, businesses, and healthcare facilities. The confines of the Internet of Things (IoTs) environments and devices operate in a manner that possesses more issues in terms of security for various devices and applications. Until now, privacy and security concerns have been analyzed extensively in IoTs domains from various aspects like data security, information transfer security, architectural security, privacy, malware analyses, and identity management.

2.2.3   Existing Gaps in Privacy Resolutions in IoTs Networks

To effectively realize the fundamentals of IoTs, analysis of its privacy and security concerns is relevant. Most significantly, IoT has been retrieved from the present technologies, which makes it possible for users to identify its challenges. As such, it is possible to embrace new or old technologies that have been in existence over the past few decades. Past research analyses focused on the differences and similarities of the privacy concerns in IoTs, with reference to ancient IT devices. Moreover, these past evaluations settled on the security issues. The fundamental driving factors that provide the basis of the differences and similarities include networking, applications, hardware, and software. With reference to these applications, there are major similarities between privacy concerns in ancient IoTs and IT domains. Nonetheless, the major concern of IoTs remains to be the resource constraints that limit the adaptations of the novel privacy remedies of IoTs networks. Moreover, the remedies to the privacy and security concerns in IoTs necessitate the cross layers, the designs, and the optimized algorithms [7]. For example, because of the computation constraints, the IoTs devices require unique samples of optimized cryptographic and other algorithms to effectively cope up with the privacy and security concerns. Apart from that, a number of IoT devices create significant challenges for the privacy technologies.

A lot of privacy concerns are more complex since the remedies obtained cannot be considered discrete. For example, as for the privacy issues like intrusions and DDoS, there is evidence of the probability of the untrue positives that will provide remedies that are not effective on the attacks. Moreover, these positives can potentially diminish the trust of consumers hence reducing the efficiency of the potential remedies. Nonetheless, the holistic privacy and security approaches over the IoTs will render nomination from the available present security resolutions, including the advancements of novel intelligence, evolution, and robust and scalable technologies used to mitigate the privacy concerns in IoTs.

2.2.4   Machine Learning (ML) – The Remedy to IoTs Privacy Concerns

The concept of machine learning (ML) represents the smart methodologies utilized to fully optimize the performance standards that make use of the present sample data and those recorded through the present learning experience. More significantly, ML algorithms propose the model of conditions that makes use of the mathematical methods on massive information sets. ML is vital in enhancing the performance of smart devices, in terms of learning, without being programmed explicitly. All these competencies are utilized to enhance the predictions useful in the future based on novel data input. ML has been considered interdisciplinary in nature since its roots can be retrieved from a number of disciplines such as engineering and science considering artificial intelligence, data theory, optimization theory, and cognitive science, among others. ML is also used when human competencies do not use or exist based on navigation hostility where users do not have the competencies to utilize their expertise, for example, speech recognition and robotics. These competencies are also used in cases where the remedies to certain issues change anytime, such as routing in PC networks or locating malicious codes in applications and software. Moreover, this aspect is practically used in smart systems; for example, Google makes use of ML in evaluating the risks against application and mobility endpoints in operating Android systems. As such, this is vital for the identification and removal of malware from all the infected devices.

In the same way, the Amazon Company introduced its service Macie, which makes use of machine learning to categorize and sort out information that has been stored in cloud storage services. Despite the fact that machine learning typically performs well in most sectors, there are normally some real negatives and untrue positives that have been recorded in the past. In that case, ML methods require modification and guidance to effectively model the most accurate predictions. Contrary to that, deep learning (DL), which is considered to be a novel option of ML, is a model used to evaluate the accuracy of all the possible predictions. Due to the condition of self-nature of the DL model, users consider it as a more precise framework used in prediction and categorization tasks to enhance the innovation of IoTs application based on personalized and contextual assistance. Despite the fact that ancient approaches have been utilized extensively in various segments of IoTs such as protocols, services, architecture, application, resource allocation, data aggregation, analytics, and clustering, huge-scale deployment of IoTs advocates for smart, reliable, and robust networking techniques. Until now, DL and ML are considered to be the novel techniques for IoTs networks because of a number of reasons, i.e., IoTs networks propose an absolute amount of information that are necessitated by DL and ML methodologies to enhance smart competencies into systems [8]. Moreover, the effectiveness of information produced by IoTs is efficiently used with DL and ML methods that enhance IoTs systems to enable more smart and informed decisions. DL and ML are massively applied in privacy, security, and malware evaluation and during the detection of potential attacks. The DL method can also be utilized in IoTs devices to perform all the complex recognition and sensing tasks used to enhance the realization of novel services and applications and to determine the actual-time interactions between humans, the physical environment, and the intelligent device. Some of the privacy-related actual-world applications of ML are considered as follows:

- Forensics face recognition: Such as posing, occlusion, lighting, make-ups, and hairstyling.
- Feature recognition for privacy recognition: Includes the various handwriting competencies.
- Malicious code identification techniques: These competencies are used to identify the dangerous codes in software and applications.
- Distributed denial of service (DDoS) detection: This form of detection technique is used in identifying DDoS attacks that influence infrastructure via behavioral evaluation. Based on the application of DL and ML methods of IoTs application, users have realized other novel challenges. All these challenges are considered multifaceted [9].

For example, it is a pending concern to effectively formulate the best model that will be utilized in processing information from various IoTs applications. In the same way, labeling inputs information in a precise way is considered to be one of the most difficult obligations. The second challenge is evident in the utility of minimal labeling information in the process of learning. The third challenge is realized in the deployment of the models on resource-centered IoTs devices where there is a need to diminish storage or processing overhead. In the same way, novel infrastructure and actual-time applications do not have the capability to withstand anomalies formed due to the DL and ML algorithms. Thus, it is considerable to systematically analyze the privacy remedies of IoTs, which leverage the DL and ML.

2.3   The Present Literature Surveys

Presently, IoTs is extensively considered, based on existing literature sources, since a lot of research has been done concerning the various aspects of IoTs privacy. In that regard, this section provides a brief overview of the present surveys before comparing these analyses with our research. According to this research, surveys have minimally focused on ML methods utilized in IoTs. Moreover, the present surveys either are based on applications or do not completely consider the complete spectrum of privacy and security concerns in IoTs networks. The present literature surveys deal with the privacy concerns in IoTs through the analysis of the available ancient remedies based on the emergent technological initiatives. Nevertheless, surveys discussing the DL- or ML-centered remedies are still not available. Moreover, DL and ML have been considered in a number of research analyses, but the general data concerning the usage of these two methods is still scarce. As for the present gap in research, we have done an analysis of the detailed survey of the DL and ML methods, which are utilized in IoTs privacy.

2.3.1   Survey Scope and Research Contributions

In this chapter, we have done a detailed systematic analysis of DL- and ML-centered privacy remedies of IoTs. Initially, we had evaluated the privacy requirements of applications in IoTs networks against threats and attacks. Thereafter, we focused on the obligation of DL and ML in IoTs, which includes the analysis of the various DL and ML methods used to leverage the IoTs services and applications. To entirely focus on the practical element of IoTs, we have done an extensive evaluation of DL- and ML-centered privacy remedies in IoTs. This evaluation also includes the analysis of the present challenges in literature and future concerns that focus on the analyses of DL and ML for IoTs networks. The main purpose for this is filling the available gaps between the necessities of IoTs privacy and the capacities of DL and ML that will permit the process of addressing the privacy concerns on IoTs networks. Thus, we have practically evaluated the IoTs from the perspective of DL and ML [10]. This evaluation focuses on privacy and security concerns of the IoTs networks. Typically, this chapter includes a detailed evaluation of the security problems and risk frameworks in IoTs. The analysis encompasses the privacy requirements and threat surfaces of IoTs, which includes the analysis of DL- and ML-centered remedies in mitigating the possible privacy attacks in IoTs networks. As such, it is considerable to argue out that the value of the present surveys is still considered to date. Moreover, the research analysis includes the recently done works in the fields of DL and ML for IoTs technologies.

2.4   Security Problems and Risk Frameworks in IoTs

Essentially, IoTs utilizes the transformational approach to effectively provide users with a lot of services and applications. This form of pervasive deployment of an extensive number of network devices fundamentally advances the degree of threat surface. Moreover, the idea that IoTs devices are normally resource-based is unfeasible to utilize novel privacy techniques to mitigate infamous threats. Moreover, it is critical to detail that the initial Internet had not been created for the IoTs. As such, it is significant to avail IoTs privacy to the present security techniques. Until the present day, IoTs makes use of various communication competencies like ZigBee, IPv6, Bluetooth, 6LoWPAN, Wi-Fi, Z-Wave, and near-field communication (NFC) [11]. All these communication technological advancements possess their challenges and shortcomings that have been borrowed from IoTs domains. Moreover, other problems include that of IP- and TCP-centered information transfer, which is prone to issues like complexity, scalability, configuration, addressing techniques, and the use of insufficient resources that pose limitations on how heterogeneous and diverse networks are applied on IoTs. Until now, various alternative technological advancements like data-centric networks and the software-defined networks (SDNs) are applicable in serving the dominant information transfer infrastructure for IoTs. In this manner, this research provides a brief evaluation of the attacks and threats that IoTs is facing. Eliminating the losses of generality, privacy concerns of IoTs can categorically be divided into various attacks as evaluated below. These threats include the following:

2.4.1   Physical Threats

When evaluating the physical threats, the intruders are provided with a direct accessibility to potentially manipulate various aspects of networking devices. To fully access all the physical devices, social engineers are required to propose the most effective methods that the attackers will utilize to access the network devices or undertake an actual-time attack ranging from physical damages on networking devices to side-channeling, eavesdropping, and many other potential attacks. Irrespective of the various technologies being used at the physical segment of IoTs, the condition of physical threats principally is similar to the requirements of social engineering and their potential approaches. Moreover, to effectively launch physical threat, attackers have to avail themselves in the close proximities of hardware and other devices with various intentions such as facilitating physical damage of hardware, limiting the devices' lifetime, endangering information transfer techniques, and influencing the energy resources.

   Physical attacks can be considered as a foundation for other potential threats such as tampering with alarms in homes, which lead to burglary and other serious damages in intelligent home environments. In the same way, the replacement of sensors with any vulnerable sensors can potentially result to the leakage of sensitive information. The incorporation of vulnerable nodes into networking devices can potentially lead to any attacks hence opening doors for other intruders to take advantage of the privileges and facilitating a dangerous attack. Moreover, this form of tampering with the security of devices can be a major facilitator of potential attacks proposing changes in security keys and routing tables that have an effect in communication to the upper degree. Other possible physical threats include jamming radio-frequencies that deny the aspect of information transfer in IoTs environments. Adding to the many other consequences, jamming leads to the denial of consumer services for the IoTs hence tampering with the operation of IoTs applications. It has therefore been considered that the intruders make use of various social engineering techniques to pose physical accessibility to the devices and hardware due to a number of reasons, which include the potential attacks mentioned above. With social engineering, intruders can possibly manipulate users to embrace physical control of the networking devices.

In IoTs environments, various issues can be created using the techniques embraced in this research. All these applications range from networking to systems, intelligent urban environments, and intelligent grids, among others [12]. Focusing on modeling, it is fundamental for consumers of technologies to make use of effective learning concepts during the initial stages. The most fundamental selection method can be divided into:

- Power-use-centered selection
- Functional-centered selection

As for the functional-centered selection method, users are given the chance to select the best concept with reference to functional variations. For instance, RL advantages from the iterative ecological perspective include agent interactive properties that require interaction with the ecology and can be used in different applications and smart systems such as intelligent temperature control frameworks and cold issues. The TML algorithm is effective for the purposes of modeling structural information, based on the highest degree of semantic character, mostly when interpretability is needed [13]. The DL-based models are normally utilized to frame out the complex unstructured information such as audios, time-series data, and images. They are the best selection criteria since big data is considered with minimal requirement concerning interpretability.

On the other hand, the power-use-centered selection method focuses on selecting the best model that proposes constraints in computation latency and power. Contrary to TML, DL and RL are typically categorized as extreme computational expenses since they normally compose the complex networking structures; hence, the accuracy determinants seem to exceed the TML based on computational overhead costs. With reference to TML at its inference stage, ideal accuracy can be attained with the most effective characteristics such as extreme-level attributes obtained from features of engineering.

1. Link Layer and Physical Privacy Problems

IoTs includes different information transfer technologies at a minimal dimension of IP and TCP protocol stacks and therefore gives more complex heterogeneous network. These networking advancements include WSN, ZigBee, Wi-Fi, MANET, NFC, and RFID. Moreover, these technologies possess their own privacy problems. This part focuses on the privacy problems in data and physical link layers of IoTs. According to literature gaps identified in this article, privacy problems in IoTs at various layers and their remedies have been evaluated. As analyzed before, heterogeneity has been proposed at its physical layers in IoTs, which further proposes various amendments in information links. For example, key channel designs will depend on the prevalent physical layer technologies [14]. Until now, security technologies of IoTs have to incorporate the heterogeneity at its data and physical link layers. There are various security concerns in physical layers of IoTs depending on the prevalent technologies; for example, in the case of sensor nodes, the attacks on these nodes have to be mitigated. Moreover, the identification of any malfunctions in hardware is a fundamental element that has to be handled with care to eliminate any form of anomaly in the upper network layer. Another privacy problem is intrusion and this requires effective countermeasures made from prevention and detection standpoints. It is significant to consider that there exist a lot of attack vectors seen in intrusions at an upper segment, for example, during the routing threats.

Based on fault detection standpoints, it is vital to identify any fault nodes in IoTs since they critically influence the quality of services (QoS) of the IoTs applications. The core purpose of IoTs is to present a ground for a minimal-energy-constrained thing (device) that co-exists at the minimal layer that assures the same information transfer for heterogeneous devices. Until now, IEEE introduced a guideline referred to as the IEEE 802.15, which allows a constrained device to transfer information in the best manner. In IoTs, the high layers utilize minimal power standards such as the Constrained Application Protocol (CoAP) and 6LoWPAN, since this is the mechanism required at a minimal layer that enables the guidelines to operate seamlessly. In this case, IEEE 802.15 gives the required changes at a minimal layer that represent the recommended standards. It is fundamental to consider that IEEE 802.15.4 includes the security concerns defining the information link layers.

Information link or MAC layers are obliged for enhancing channel accessibility for various devices in addition to access management, framework validation, security, and time management. In this research, we have focused on privacy problems in the high layer given by IEEE 802.15.4. These securities given by the standards at MAC layers ensure that the levels of the nodes and information transfer are kept secure. Moreover, these securities ensure that the securities in the upper layer are maintained. In that case, symmetric cryptographic algorithms like the AES are recommended to be implemented efficiently and rapidly on chips; hence, all these implementations in IEEE 802.15.4 hardware shall be considered in the lower layer privacy. The IEEE 802.15.4 standard provides the AES algorithms and various applications that are recommended to control the resource constraints of the networking devices. The guideline supports various privacy nodes at the link layers; for example, information might not be encrypted by just executing an integrity check.

2.4.2   Networking Layer Threats

In the networking level, the threats are focused on routing, traffic evaluation, man-in-the-middle, and spoofing attacks. Apart from that, Sybil influences are potentially seen in the layers of the network where false Sybil identities are utilized in creating illusions in networking. Intrusions via various means propose a manner in which the intruders of the networking system can potentially launch some possible threats. As such, ensuring the safety of networks is fundamental to deal with the threats at their initial stages. At the networking layers, the intruders have possibly leveraged the vulnerable nodes to perceive them as fake-forwarding nodes facilitating the formation of sink holes.

This form of threat normally is linked with the mobile ad hoc and sensor networks, which pose a significant effect on IoTs environment. With all these threats, there is a chance for launching an associated DDoS threat that will affect the entire IoTs network. In the networking layer, the intruders can potentially affect the network by bombarding it with a lot of traffic that comes from the compromised nodes beyond what their network can possibly handle. Affecting the IoTs nodes and tampering with the identities will pose a significant effect on networks since there are fake nodes that allow these intruders to prepare Sybil threats where the Sybil nodes provide an illusion to the major networks as though the actual nodes were transferring the data sets [15]. To draw the general assumption of the threat vectors in the networking layer, users should focus on the information transfer aspects of IoTs and utilize the resource constraints, authorization frameworks, and sophisticated authentication.

2.4.3   Transportation Layer Threats

The transportation layers are tasked with the obligation to enhance step-by-step delivery of transportation standards that enhance the procedure followed during the exchange of information. In this case, the ancient transportation layer privacy concern typically persists. The most serious threat in this layer is the denial of service, which affects the network applications. Moreover, it is critical to note that due to the status of IoTs, UDP and TCP standards have no scale with the resource-constrained device, and hence, the lightweight version of transportation guidelines had been recommended in research. Nonetheless, the privacy problem of the standards is of a major concern that effectively enhances the DDoS and DoS threats in IoTs.

2.4.4 Application Layer Threats

The IoTs applications are considerably the best target for the intruders since attacks on the application layer are comparatively easier to launch. The most known threats include buffer-overflow threats, denial of service, malware threats, phishing, cryptographic threats, exploitative web app threats, man-in-the-middle, and side-channel threats. The buffer-overflow attacks are considered to be the widely utilized threat vector in various applications [16]. The present methods used to deal with the buffer-overflow attacks include dynamic nodes and static nodes analyses, which include other complex techniques such as symbolic debugging techniques. Nonetheless, the methods cannot be utilized with IoTs because of resource scarcity. The IoTs applications are also affected by malicious code incorporation due to buffer-overflow attacks, in addition to vulnerabilities like cross-site scripting, SQL injection, and object referencing, among others.

According to the Open Web Application Security Project (OWASP), there are a lot of susceptibilities that lead to various threats on network applications. One of the latest susceptibility that OWASP recorded was in 2017. The susceptibility produces a collection of a lot of threats, which can be launched by attackers at any time. For example, attackers can choose to inject malicious nodes, access controls, perform phishing, and tamper with privilege escalation. Moreover, using malicious code injection, the intruders can possibly collect sensitive data, tamper with the data sets, and do a great deal of malicious activities. Botnets pose another fundamental threat to the IoTs application and infrastructure. Controlling the threats produced via smart botnets poses a significant problem for the IoTs due to the fact that these botnets effectively crawl and scan the networks searching for unknown vulnerabilities before exploiting them to enhance the launching of massive DDoS. It is significant to mention that because of the scarcity of resources, the modern cryptographic standards are unfeasible for IoTs devices that expose them to potential intruders who launch the cryptographic attacks. Generally, these threats are seen on the application layers of the IoTs infrastructure, which makes it much costly to control.

2.4.5 Multiple-Layer Threats

The multiple-layer threats add to the aforementioned threats in this chapter and are launched to the IoTs infrastructure. These threats include side-channel, traffic evaluation, man-in-the-middle, relay, and standard threats. A lot of these threats have been evaluated in the section above. The traffic evaluation threat is considered as an attack where the intruders control the traffic and make use of it. Users find it difficult to control this threat since communication parties normally have no knowledge that their networks are being monitored. The intruders are searching for vital data sets in the Internet traffic; the data include personal details, company logic information, and credentials, among others. The idea of information transfer privacy is of great significance to IoTs. The information that is produced in the IoTs ecology is utilized for the purpose of decision-making. Hence, it is fundamental to ensure the health and quality of information. Tampering with the privacy of information in IoTs will pose a significant challenge on the vulnerable applications. It is also fundamental to mention that SDN has been significantly leveraged to attain a wide speculation of advantages from the IoTs security and applications [17]. The novel functionalities given by the SDN control plans enhance companies to significantly control a lot of things and sensors in the IoTs paradigm. Nonetheless, irrespective of the SDN merits that have been mentioned above, the available interface in SDN provides risk to threats on the vulnerable IoTs infrastructure, applications, and devices. In that regard, the privacy concern of the IoTs is dependent of the privacy of the SDNs.

2.5 Machine Learning and the IoT Safeguarding

This section discusses about different machine algorithms alongside their usefulness within the IoT applications. First, the section shall commence with the study of machine learning which can be segmented into four major sections, which include supervised, semi-supervised, unsupervised, and reinforcement machine learning.

Supervised machine learning: This type of learning is executed if accurate targeted points have been prospected to be obtained from a specific input set. Therefore, based on this system of learning, there is labeling of the data, which comes first, and then the training of the marked data sets. These types of sets voluntarily rule from the availed sets of data and furthermore give meaning to different types of classes and in the end foresee the elements that fall under a prospective class. On the other hand, supervised learning algorithms may be utilized if data X aligns itself with data Y, which are provided for the training whose main objective is to adjust to a mapping function which is Y: $\leftarrow$ f (X). There has been extensive application of supervised learning algorithms within the IoT sector; for these reasons, there should be an introduction of various classifiers as shown in this chapter. Both logistic regression and perception can be regarded to be the easiest linear classifiers. Hence, for the two frameworks, they can be understood to be the simplest linear transformations. Whereas perception may execute binary categorization centered on the signal that the input data is being transitioned, RL can increase the rate of the transformation to a specific probable value, prior to getting a threshold function, which will be applied when making the classification decision [18]. Moreover, the RL may also be increased to multi-class classification events with the aid of the softmax function, which is a scaling function that has class-related likelihood to the primary output. The artificial neural networks (ANN) can simply be as previously mentioned linear classifiers. In contrast to the perception or rather the RL that is a linear-related project input data which is linked to the output, the ANN comprises extra "secretive layers" that permits the ANN for modeling non-linearity.

However, in relation to the linear classifiers, it is the additional secretive layer that leads it to become rarely known to see any linked association amid the output and input data sets. However, in theoretical basis, having one hidden layer within the ANN model will lead to the modeling of non-linear functions, in relation to the restricted capacities during the encounter with unforeseen data sets. Moreover, the ANN that has more layers will be known to be a deep neural network, which may seem to have improved modeling capacities as mentioned in the following sections.

Unsupervised machine learning: Regarding unsupervised learning, the vicinity normally issues the available inputs deprived of the targets desired. However, there is no need for a labeled data set which can analyze the uniqueness amid the unlabeled data and categorize the data into dissimilar sections. The supervised and unsupervised learning tactics specifically aim at the evaluation of different data set challenges, whereas reinforcement learning can be utilized for decision-making processes and various comparisons. Therefore the grouping and choice selection of the ML methods relies on the manner at which the data can be accessible. If the input data and prospected output are known, supervised learning can be employed. During these scenarios, the entire framework comprises only trainees so that the input can be mapped appropriately and a stipulated input can be achieved. Regression and classification are various instances of the supervised techniques whereby regression operates consistently while classification works alongside discrete outputs. There are different regression methods such as supported vector regression (SVR), polynomial regression, and linear regression, which are the mostly utilized methods. Subsequently, the grouping operates with various discrete output figures.

This is mostly used in instances for grouping various algorithms such as the k-nearest neighbor, SVM, and logistic regression. Other algorithms may be utilized for the grouping and classification, for example, the neural networks. Outputs have not been properly defined and clustered with various classifier objects centered on known criteria for example k-clustering [19]. The extent of accuracy for the foreseen analytics is based on how enough the ML methods have utilized recent data for creating new models and the way they have been used for predicting various models. Various algorithms, for example, the SVR, the naive Bayes, and the neural networks, have been utilized for predictive frameworks.

2.5.1 The Main Objective

The main objective of the unsupervised learning algorithms is to enhance the understanding of the web relationship amid information that comprises only X data, which is only present when the class Y is not available. For instance, algorithm clustering may be utilized in order to identify various capable series with others that have not been named and acquired outcomes that can be utilized for further evaluation. Therefore, k implies to be the core component analysis, which falls among the recognized unsupervised algorithms. With k, it means that the main objective is identifying a group series amid information through assigning various samples of clusters centered on the range existing between the centroid and the samples based on every cluster. On the other hand, the PCA can be dimensionally utilized as a reduction that can relate with various raw elements prior to choosing ones that are informative.

The semi-supervised learning: Comparing the initially mentioned groups, there will be no tags that exist for the observation of the data sets or rather the labels, which have presented all notable observations. Considering the vast number of practical scenarios, the expenses used in labeling can be very high; hence, it needs equipped and talented human resource to be employed. In reinforcement learning (RL), having no accurate results can be explained, and every agent can understand from the response once it has interacted carefully with the neighboring surroundings. For this reason, it executes various actions that lead to the implementation of various decisions based on the reward acquired [20]. Therefore, for an agent which has been rewarded for the execution of recommendable actions or castigation of poor actions and utilize feedback methods so that they can increase long-term recompenses. This has been immensely inspired through the learning of various character traits of animals and people. These behaviors have established a pretty approach within the dynamic use of bots whereby different systems may learn to accomplish various roles by not using explicit programming. It is crucial that people should select an excellent reward so that the triumph and catastrophe of these agents may rely on various gathered rewards.

For this segment, there will be an introduction of various techniques that can be utilized in formulation of the previously mentioned streaming video example that comprises RL. Considering the previously mentioned results, RL, which is an agent that associates itself with the neighboring environment, to get a customized control policy, can be gained through experience. Therefore, there is a need for involvement of three basic elements, which include action, reward, and observation. Centered on the mentioned elements, formulation of the adjustable bitrate streaming challenges can be easily possible. Uniquely, various observations may buffer various occupancies, the network throughput, and so on. For each phase, every agent may decide on the bitrate of the incoming chunk. Hence, a reward can be acquired once an agent receives the response. The algorithm proposes, gathers, and then simplifies the outcomes used for executing the initial decisions; hence, it can optimize various policies based on various network coverages. The RL based algorithms may lead to environmental based noises such as unforeseen networking situations, video assets etc. Various DL architectural features that can be accessible in literature may include the CNN (convolutional neural networks), BM (Boltzmann machine), DBN (deep belief networks), RNN (recurrent neural networks), FDN (feed-forward deep networks), GAN (generative adversarial networks), and LSTM (long-short-term memory). For the CNN and the RNN, they are extensively utilized for deep architectural learning [21].

Deep reinforcement learning: In this segment the DL can be a type of ML method that is utilized for estimating the function, grouping, and foreseeing while the RL will be a different type of ML technique which can be employed in decision-making processes whereby every software may be educated pertaining to how various optimal actions can be achieved though the interaction with the nearby environment with some states. Both the RL and the DL chip in within these circumstances and the data is dimensionally huge while the environment can be non-stationary. Hence, conventional RL will not be sufficient. Through the combination of both the DL and the RL, various agents may learn on their own and draft effective policies in order to extract maximum substantial rewards. Considering this approach, the RL receives aid from the DL in order to allocate an excellent policy while the DL executes various action estimations so that it can get the quality of actions within a prospective state. Moreover, the RLL and the DL gain advantages from one another [22]. The DL is in a position of understanding the more complicated series although it is exposed to various inaccurate groupings. For this situation, the RL comprises a powerful capacity for automatically understanding the environment by not constructing and aiding the DL during effective grouping.

2.5.2   IoT Security Machine Learning Methods

In this section, there will be a discussion of some ML algorithms that are specifically aimed at underlying safety and confidential challenges within the IoT networking system. Accurately, there will be consideration of various authentic invasion techniques and other mitigation practices, DDoS threats, intrusion sensing, anomaly, and some malware evaluations. The supervised learning algorithms operate using tagged information and use an IoT network within spectrum sensing, adaptive cleaning, channel approximation, and localization challenges. Such a section comprises two speculated forms of methods, which include regression and grouping. Grouping within supervised machine learning can be utilized in foreseeing and also in modeling accessible sets of data. On the other hand, regression can be used in predicting consistent patterns of variables. Some few widely employed classifications of algorithms include the naïve Bayes, decision tree, SVM, and random forest [23]. Therefore, the SVM employs a framework known as kernel, which is utilized in identifying various disparities between two common points that comprise different classes.

The SVMs are in a position of modeling the non-linear sections. The SVM can genetically include the memory and hence be hard deciding the effective kernel, there by modeling large sets of data becomes difficult. Hence, a random forest is usually recommended unlike the SVM. On the other hand, the naïve Bayes (NB) has been using Dover time in helping solve global problems such as the spam sensing and classification of text. Considering having all naïve and other input elements, every other random forest provides an ideal environment for the modeling of actual problems. With random forest algorithms, it will be much easier to implement and adjust huge-sized data sets. Such algorithms may take a much longer period of practicing compared to the supervised algorithms, which include the NB and the SVM. However, they can attain a greater accuracy and may consume less amount of time in foreseeing various issues. Moreover, they have been centered on the structuring of various graphs that segment into leafs and branches, hence showing the class and decision that have been implemented.

2.6 An Analysis of Present Machine Learning Centered on Solutions Within the IoT Security

This section provides a quick survey of the present ML-centered solutions that highlight dissimilar safety problems within the IoT. Authentication and access within the IoT system has been a primary requirement for users within the IoT. Every user should be authenticated so that every application or service can be utilized effectively. Normally, these applications can focus on the exchange of data within different conditions. Such data can be accessed within the IoT devices and it is processed and directed to the decision-aided system in order to acquire accurate meaning from it. Such processes can range based on the architecture of the IoT although the flow of information can be the same within these frameworks. Having no loss in generality, if a certain application or any client requires some data within the IoT gadgets, then the entity should be effectively authenticated [24]. If this happens, the request of accessing the information can be neglected. Similar to other networks, the access control can be of significant use within the IoT networking system and similarly can be problematic considering that it can be restricted by network volume, heterogeneity, asset limitation of various devices, security of various networks, vulnerable threats, etc. Moreover, it is crucial for giving and denying various clients access to critical information for the applications that use the IoT services. As discussed regarding the ML-centered mechanisms within the IoT, first it is essential to focus on various segments that have access control.

According to scientific research, the access control can be divided into different sections such as CWAC (context-aware access control), RBAC (role-based access control), and PBAC (policy-based access control). Other researches have shown that improved groupings include the ABAC (attribute-based access control), CAC (capability-based access control), UCAC (usage control-based access control), and OAC (organization-based access control). Some of the present works may be seen within the mentioned sections. On the other hand, a comprehensive survey was carried out and utilized the access control mechanism within the IoT network hence pointing out various pitfalls within the IoT. Various authors may analyze the present mechanism centered on the outstanding applications for the IoT. Such applications are widely segmented in primary classes, the enterprise and personal. There are a lot of applications such as healthcare, minor offices, digital homes, sensor networking systems, smart firms, perilous infrastructure, business applications, and many more. The access control systems are utilized within applications and within the architectural levels.

For the architectural layers, every product or service comprises a collection of decisions which include the access control languages such as XACML (extensible access control mark-up language), UMA (user-managed access), and OAuth (open authorization). Such architectural phase control contrivances can be utilized by using the present protocols within the IoT network; for example, the OAuth can be enacted across the present IoT protocols, for example, the CoAP (constrained application protocol) and the MQTT2. Additionally, the various services provided by Facebook, Google, Netflix, Microsoft, and many others comprise many client accounts, which all utilize OAuth [25]. Subsequently, a discussion pertaining to the present DL- and ML-centered authentication and various accessible mechanisms within the IoT networking systems is presented below.

Access control for ML-based authentication processes within the IoT: According to extensive analysis, a physical layer of authentication mechanism has been suggested within the IoT network. Hence, the suggested mechanism utilizes various physical properties, which include the strength of every signal. The essence is that some revolutionary techniques have been used for physical layer verification processes and gaming theoretical approaches, and these techniques are machine-based, which ensures that there is effective isolation of various spoofers which emanate from the benign IoT clients. The authentication systems can be created spoofing method that can be attempt the utility while increasing the threat frequency. On the other hand, the frequency channels get feedback, which is utilized in establishing a Nash equilibrium (NE). Hence, it is important to point out that every packet acquired via the radio channels comprises various channel states that can leverage the test threshold focusing on the authentication of every decision, which would have been created. Due to this reason, various scholars have attempted to use Q-learning and Dyna-Q that aid in comprehending the state of every channel while not acquiring enough data pertaining to the channel. With the aid of experiments, scientists have concluded that sensing precision and execution process for the Dyna-Q model is effective compared to the Q-learning methods. A familiar physical authentication layer project within the distributed Frank-Wolfe (dFW) algorithm has been recommended.

The access control of the DL-centered authentication process within the IoT: The prospected client's authentication methods used within the IoT focusing on the physiological events caused by human beings have been influenced via Wi-Fi signals. Therefore, the recommended scheme has been based on various recognition events for humans. Activities may be executed using data that are coarse-grained and have features that are smaller. Due to this reason, the information states for every channel within the Wi-Fi signals have been created by the IoT gadgets, and hence these characteristics are so much dissimilar considering their features. The DNN (deep neural network) can be used in learning more about human physiological and various behavioral qualities, which can be useful within the authentication process. For the three layers, there will be extraction of different forms of events, and at the second layer, there will be learning of the activity, while the third layer will comprise highly based elements that are centered on the authenticity of each user.

2.6.1 Mitigation and Detection of Threats

Following the subsequent sections, there was concise explanation of various invasions that were launched within several layers of the IoT network. Due to the limited number of resources and heterogeneity for the IoT devices, they create a sufficient platform for threats to come in. Normally, attackers may use the identified vulnerabilities within all the networks, and the gadgets can be executed using dissimilar types of invasions. Therefore it is recommendable to mention that the invasions may range from low-profile hacking to a gadget that is enormous, for example, the WannaCry, and can be much more complicated such as that of Dyn and Mirai. Conventional invasion sensing and various mitigation systems can be centered on cryptography, and at times there can be inadequate precision, and this may result to inaccurate positives. For this reason the ML-centered methods, for example, the DL, KNN, and SVM, and the unsupervised learning methods can be used. In this section, there will be a brief outline of the ML-based techniques that are used for attack sensing and various mitigation processes within the IoT networking system. Moreover, in this section there will be an analysis of present DL and ML techniques.

The ML invasion sensing and its mitigation used within the IoT network: There has been proposition of the semi-supervised attack sensing mechanism that can be used within the IoT. In real sense, the recommended system is centered on the use of the ELM (extreme learning machine) algorithms alongside the FCM (fuzzy C-means) techniques, generally referred to as the ESFCM. The ESFCM can be enacted within the fog substructure. One of the speculated features of the ESFCM is that it deals with tagged information hence augmenting the sensing rate within the evenly spread attacks. Although the sensing precision within the ESFCM can be lower compared to the initially mentioned DL mechanisms, it may outwit the conventional mentioned ML algorithms used for sensing various invasions [26]. Nonetheless, some semi-supervised learning systems draw various elements from both the unsupervised and supervised learning systems, which may make them effective compared to other systems. As stated previously, the IoT comprises a vast number of breeds in which some of them include personal networks and may become more complex within the infrastructural industry, for example, the smart grid.

The DL-centered invasion sensing and mitigation techniques within the IoT: The sensing of the DL-based invasions within the IoT through leveraging of the fog system has been stated. For this reason, the invasion sensing system can be enacted within the edge of the smart infrastructure. Hence, the aligned attack sensing mechanism comprises various account parameters that include the learning systems and will choose the type of output that will be allocated to certain data. Hence, the theory that has been employed in utilizing the fog technology is an effective asset limitation and a natural application within the IoT. In an event where there is intense infrastructure, therefore the learning mechanism can be close to the information nodes so that there can be timely delivery and quality decisions can be made when there is an attack.

2.6.2   Attacks Within the DoS and Distributed DoS (DDoS)

DoS and DDoS are the two most dangerous invasions whose attacks are always difficult to alleviate within the IoT surroundings. Many claims have been stipulated so that enough reasons can be provided to satisfy the question of why these attacks are immense. Some of the reasons may include a sheer number of linked IoT gadgets being heterogeneous, the Internet, insufficient mechanisms considering the limitation of resources for the IoT gadgets, communication within platforms, extensive communication, etc. Such scenario illuminates various dangers for the IoT gadgets in contradiction to the DDoS invasions. These types of attacks can be evaded so that there can be assurance of sufficient operation of the IoT applications. Based on this, the IoT can be known as a land of chances applicable to the DDoS invaders. During 2016, there was an extraordinary increase in invasions in contradiction to the IoT substructure extensively. An example of such an attack was Mirai4 that led to the decline of the Internet whereby devices, for example, printers, babycams, etc., have been utilized like robots for the execution of the DDoS invasions upon many firms [27]. Equally, some like Mirai bots were reported also. With its inception in 2016, the malware led to substantial disruption within Internet services based on its complicated contagious contrivance within the IoT networking system. Due to the claim that only IoT devices can be hacked, they can be utilized and be regarded as front tools for the execution of disreputable invasions in contrast to various organizations, which advocate intense intellectual safety measures, which can safeguard their equipment.

Until now, crucial research outcomes have been acquired via different methods in order to alleviate the DDoS threats within the IoT, although dissimilar architectures ensure that it is hard to create a combined system in order to curb the DDoS threat within dissimilar IoT podiums. Conventional DDoS sensing and vindication systems within the IoT networking system have been used on routers, gateways, or accessible points with the aid of intrusion sensing and anticipation methods. Based on the previously mentioned protocols, both the CoAP and the MQTT can be utilized as protocols for the IoT. According to scientific research, the DDoS invasion has been evaluated based on the attack on the generic IoT, which utilizes the CoAP. Such works can be utilized to simulate an invasion in order to evaluate the capacities within the IoT network. Additionally, in order to sense and mitigate the DDoS attacks, some enabling technologies, for example, fog and cloud computing, can be utilized for assisting the detection of DDoS systems within the IoT. For example, fog computing can be used as a centered approach for safeguarding various malicious threats. Amid several flavors for the IoT networking systems, various critical development-assisted networks should portray rigid resilience for curbing the DDoS invasions.

Cloud computing and fog DDoS mitigation frameworks can also be suggested. Within the DDoS levels, a conventional mechanism can be employed within several layers within the IoT industrial networking systems. Hence, the minimum levels that utilize the computing edge along with the SDN gateways and the networking traffic can be accumulated within a fog computing level that comprises the DDoS. Moreover, the honeypots can be used within this phase. At last, the cloud computing levels can be created and evaluated within a cloud-based platform in order to sense various vulnerable DDoS invasions. Based on the previous studies, it is vividly clear that there will be no bullet, which is in a position of catering for the DDoS threat within asset-restricted network coverage, and it is known that literature that is intellectual can be important in sensing and mitigating the DDoS. Moreover, falsified values cannot be questioned in these scenarios; hence, the request can be neglected [28]. For these reasons the asset-limited and insufficiently effective gadgets within the IoT networking system may bring out an alluring environment that can be good for harmful threats. In spite of the latest developments in understanding such types of invasions, still it is important to work on the intellectual contrivances that do not only accompany the traffic amount but are also characters of the threat. In considering the subject of this issue, machine learning can be regarded as an efficient candidate that can be leveraged for use in DDoS sensing within the IoT networking systems. In order to alleviate the DDoS invasions within the IoT networks, the research sector has been in a position of leveraging some ML and DL experts.

1. The ML-centered methods to highlight the DDoS and the DoS invasions within the IoT: First, there is a detailed rationale based on the present ML systems for the DDoS sensing within the IoT networking systems. The typical elements within the IoT systems in which involved consistent communication lacking the back end networking servers. Moreover, there was a comparison between the decision tree, the k-nearest, the random forest, the neural networks, and the SVM based on whether they can sense DDoS in the IoT. On the same note, within the SDN environment, the SVM-centered DDoS sensing systems have been used. Apart from this, for DDoS sensing, the SVM has been proved to have high accuracy compared to other mechanisms. Other techniques can also be compared with others such as the RBF (radial basis function), naïve Bayes, bagging, and random forest.

2. The ML-based IDS within the IoT: There was a proposition of using ML-based IDS that is lightweight, has low power, and can be used for operating the 6LoWPAN. This mechanism utilizes three methods such as the K-means, the decision tree, and the hybrid methods which combine the previously mentioned methods. Furthermore, the leveraged ML methods can sense the intrusions within the IoT pathways.

3. The IDS within the IoT heterogeneous networking systems: Within a heterogeneous networking system for example, the RNN (recurrent neural network) has been utilized within the LSTM infrastructure. RNN (random neural network) has also been used that can be effective for the realization of the fast intrusion-based sensing within minimal power of the IoT networking system.

2.6.3   Analysis of Malicious Programs Within the IoT

The depreciation in number alongside the heterogeneity of IoT gadgets has provided a cool and excellent platform for cybercrimes. The potential vulnerability can be utilized within the system. A notorious invasion within the domains can be done through vulnerable code injection which can affect the performance of the present IoT gadgets. The threat, which can be performed by injection of different malware, can be linked to the use of various authorizations, security protocols, and authenticities. Aside from the mentioned techniques altering the IoT gadgets so that the software can be physically modified can aid attackers inject vulnerable codes. Prior to involving yourself into deep data based on the malware, it is crucial to comprehend various groups of malicious programs that may put the IoT security at risk. Generally, a malware can be a persistent threat based on the previously mentioned vulnerabilities that can be performed via various invasions. Common forms of malware include, although not restricted to, ransomware, bot, Trojan, adware, etc.

This section will summarize the groupings of different malware, which influence the operation of the IoT gadgets, and further analyze the present solutions such as ML-centered methods, which maintain the IoT gadgets' security. According to scientific study, a vast number of smart gadgets that have been linked to the Internet are deprived of enough security, which not only is dangerous to the devices but also may permit the invader to gather resources meant for enormous attacks, for example, the DDoS. For example, scientists have attempted to test different music gadgets for various problems and used vulnerable codes to show that these devices are highly prone to different attacks when linked to the web.

Invaders also can utilize Internet-based webcams installed in public places, restaurants, and residential places for threatening intentions. Furthermore different types of malware have been created for the disintegration of normal operation of a business and other entities; these malware programs include, although not restricted to, the Night Dragon, NotPetya, CryptoLocker, Stuxnet, etc. There are detailed classifications and generalized principles for various malware programs. Such may include the generic attacks but there are also customized families for the malware threats which are specifically aimed at the IoT gadgets. Many malware invasions may include CryptoLocker, WannaCry, Stuxnet, etc. Such attacks have caused substantial losses for the industry and depreciation of public image for the entire firm. Information connected to these threats is all over the entire chapter. It is crucial that people comprehend a generalized concept for the invaders for launching the threat using a malware.

Precisely, the invaders may accumulate ideas pertaining to capable targets, for example, various network sensors via taking a reconnaissance. A vast number of methods are present which can be used for executing a reconnaissance, such as Wireshark, Nmap, and Metasploit, alongside the use of social engineering techniques. Various present tools give a chance of additional data based on the different uses that may create a more conducive platform for threats to come in. For this reason, every attacker will first think about the type of threat he/she would use in

relation to the uniqueness of the device class. Until now, certain application evaluation techniques, for example, the OWASP (Open Web Application Security Project),
give access to the primary sources of threats by attackers making them to select an
effective exposure [29]. Generally, the OWASP gives an opportunity to various
sources so that they can exploit the resources that can be utilized by various invaders, for example, the injection of SQL, misconfiguration of security measures, mis-
guided authentication, and XSS (cross-site scripting).

Moreover, based on the device type and its exploitation, the attacker can transmit
a payload to a specific target via the use of various media, for example, rootkit,
phishing, and various updates. For these reasons, the malware group can range from
a simple sole malware to a multifaceted, intellectual, inactive, and versatile emanat-
ing malware. Currently, intelligent malware can be very adaptive based on the vicinity of the IoT whereby it can be used for getting access to the prospected network
and adjusting the performance processes based on the network. For example, various malware may go astray when detected and can be latent for a certain period;
hence, they do not perform malevolent coding during that period so that their main
intention would not be compromised. Based on this concept, various measures that
can be used for evading the detection method are as follows:

1. Malware elusion methods: Different techniques that are used for evading these
   malware can be used. Encoded malware implies to the encrypted harmful codes
   in relation to the subsequent decryption so that it may surpass the signature-
   linked antivirus. Although the decryptor tends to be the same, for dissimilar version in relation to the same malware, it still remains to be detected. In order to
   overwhelm all of these, a decryptor can be transformed and hence will avoid the
   detection contrivance. Such a form of malware is known as oligomorphic malware. Correspondingly polymorphic malware creates diverse decryptors that
   will make it hard for sensing engines to identify the malware [30]. The metamor-
   phic malware can be indeed the complicated malware among other groups since
   it emanates to a new generation whereby it can be dissimilar to any other for
   detection.
2. Malware ML-centered evaluation within the IoT: Moreover there has been
   accumulation of various supervised learning methods that use random forest
   groupings which are employed for sensing malware that is Android-based.
   Therefore every classifier's sensing precision can be evaluated based on the
   malware checklist with the application of Androids. There has been an investigation on the techniques for sensing various malware and their propagation in
   the WMS (wireless multimedia system) in IoT. Research has shown that cloud-
   based techniques within the SVM have been leveraged in order to sense various capable malware and their propagation and can be utilized in dissimilar
   gaming so that they can minimize the malware threats. Just after acquiring the
   Nash equilibrium, there has been an attempt of looking out for a conducive
   plan for the WMS so that it can protect itself against malware. Correspondingly,

the prospected SVM linear methods for the classification of various malware that are Android-based are used.

The detection system is centered on ANN before being evaluated by malware and benign application in Android devices. Essentially, MalDozer is centered on a series like API technique calls in resource permission, Android, and fresh method calls, among others. Moreover, the recommended methods include the automated engineering characteristics during the process of training. Moreover, the research proposes an image recognition-centered DDoS malware-detecting method of IoTs network. The research recommended a solution, whereby the researchers collected and categorized two significant segments of malware known as Linux and Mirai. Thereafter, research practices convert the segment binaries of the IoTs application to the grayscale image. Afterward, CNN is conditioned to categorize the images to malware and good-ware. Other researches considered using the deep auto-encoders to show the botnet threats in IoTs. In the solution identification, the researchers evaluated the behavior of networks before using the deep auto-encoders to separate any form of anomaly in the network. Thus, the discussion concerning DL and ML approaches for the detection of malware in IoTs networks indicated that the techniques used in data training are used to identify indefinite malware.

2.7   Conclusion and Future Contributions

In conclusion, the emergent advancements recorded over the past few years show that Internet standards and computer systems have enhanced the process of communication between various network devices. Approximately 25 billion devices are speculated to link up with the Internet by 2020. This speculation gives rise to the novel development idea of the IoTs, which is a combination of the embedded advancements such as wireless or wired communication devices, actuator and sensor devices, and physical objects linked to the Internet. The most vital objective of computing is to enrich and simplify human experiences and activities which is also one of the visions linked to twenty-first-century computing. The IoTs necessitates information to provide effective services to individuals or improve the IoTs system evaluation to attain a smart environment. In that way, frameworks have to be able to access raw information from various resources over the networking devices and evaluate data to extract user competencies. Big data is explained as a high-velocity, high-volume, and high-variety data set, which is expensive and innovative to enhance user automation, decision, and insight. In consideration to the challenges caused by big data, the future relies on the launching of novel concepts of intelligent data to enhance productivity, effectiveness, and efficiency.

References

1. Anandakumar, H., & Umamaheswari, K. (March 2017). Supervised machine learning techniques in cognitive radio networks during cooperative spectrum handovers. *Cluster Computing, 20*(2), 1505–1515.
2. Anandakumar, H., & Umamaheswari, K. (October 2018). A bio-inspired swarm intelligence technique for social aware cognitive radio handovers. *Computers & Electrical Engineering, 71*, 925–937. https://doi.org/10.1016/j.compeleceng.2017.09.016.
3. Adli Mehr,K., &Musevi,N. J. (2016).Security bootstrapping of mobile ad hoc networks using identity-based cryptography. *Security and Communication Networks, 9*(11),1374–1383, Available: https://doi.org/10.1002/sec.1423
4. Bredt, S. (2019). Artificial Intelligence (AI) in the financial sector—Potential and public strategies.*Frontiers in Artificial Intelligence, 2*, Available: https://doi.org/10.3389/frai.2019.00016
5. Chittaro,L.,&Ranon, R. (2004). Hierarchical model-based diagnosis based on structural abstraction. *Artificial Intelligence, 155*(1–2), 147–182. Available: https://doi.org/10.1016/j.artint.2003.06.003
6. Chatzigiannakis,I., &Nikoletseas, S. (2004) Design and analysis of an efficient communication strategy for hierarchical and highly changing ad-hoc mobile networks. *Mobile Networks and Applications, 9*(4), 319–332. Available: https://doi.org/10.1023/b:mone.0000031591.74793.52
7. Haldorai, & Kandaswamy, U. (2019). Supervised machine learning techniques in intelligent network handovers. In *EAI/Springer innovations in communication and computing* (pp. 135–154). New York: Springer. https://doi.org/10.1007/978-3-030-15416-5_7.
8. Delgadillo,G. Ó. (2016).Fishes and mobile macroinvertebrates of artificial habitats in Taganga Bay, Colombian Caribbean. *Bulletin of Marine and Coastal Research, 38*(1), Available:https://doi.org/10.25268/bimc.invemar.2009.38.1.338.
9. Dubois,D.,&Prade,H. (2001). Special issue of the journal *Artificial Intelligence* on "Fuzzy set and possibility theory-based methods in artificial intelligence."*Artificial Intelligence, 127*(2), 269–270, Available: https://doi.org/10.1016/s0004-3702(01)00080-7.
10. Dubois,D.,&Prade, H. (2003).Fuzzy set and possibility theory-based methods in artificial intelligence. *Artificial Intelligence, 148*(1–2), 1–9, Available: https://doi.org/10.1016/s0004-3702(03)00118-8.
11. Ebendt,R.&Drechsler, R. (2009).WeightedA∗search – Unifying view and application. *Artificial Intelligence, 173*(14), 1310–1342, Available: https://doi.org/10.1016/j.artint.2009.06.004.
12. Giudici, P. (2018).Fintech risk management: A research challenge for artificial intelligence in finance. *Frontiers in Artificial Intelligence, 1*. Available: https://doi.org/10.3389/frai.2018.00001.
13. Gong,T., &Bhargava, B. (2012).Immunizing mobile ad hoc networks against collaborative attacks using cooperative immune model. *Security and Communication Networks, 6*(1), 58–68, Available: https://doi.org/10.1002/sec.530.
14. Gu,B., &Chen, I. (2005).Performance analysis of location-aware mobile service proxies for reducing network cost in personal communication systems. *Mobile Networks and Applications, 10*(4), 453–463, Available: https://doi.org/10.1007/s11036-005-1557-x.
15. Guo, M. (2011).A simple and rapid authentication protocol in mobile networks. *Security and Communication Networks, 7*(12), 2596–2601, Available: https://doi.org/10.1002/sec.386.
16. Jávor,A.,&Szűcs, G. (1998).Simulation and optimization of urban traffic using AI. *Mathematics and Computers in Simulation, 46*(1), 13–21. Available: https://doi.org/10.1016/s0378-4754(97)00154-7.
17. Kharin, Y. (1994).Artificial intelligence frontiers in statistics AI and statistics III. *Knowledge-Based Systems, 7*(1), 57–58, Available: https://doi.org/10.1016/0950-7051(94)90017-5.
18. Lasi, H. (2013).Industrial intelligence – A business intelligence-based approach to enhance manufacturing engineering in industrial companies.*Procedia CIRP, 12*, 384–389, Available: https://doi.org/10.1016/j.procir.2013.09.066.

19. Leo Kumar, S. (2017).State of the art-intense review on artificial intelligence systems application in process planning and manufacturing. *Engineering Applications of Artificial Intelligence, 65*, 294–329. Available: https://doi.org/10.1016/j.engappai.2017.08.005.
20. Marquis,P.&Schwind,N. (2014).Lost in translation: Language independence in propositional logic – Application to belief change. *Artificial Intelligence, 206*, 1–24, Available: https://doi.org/10.1016/j.artint.2013.09.005.
21. Mihăiţă, A.,Dupont,L.,&Camargo,M. (2018).Multi-objective traffic signal optimization using 3D mesoscopic simulation and evolutionary algorithms. *Simulation Modelling Practice and Theory, 86*, 120–138, Available: https://doi.org/10.1016/j.simpat.2018.05.005.
22. O'Halloran,S.,&Nowaczyk, N. (2019).An artificial intelligence approach to regulating systemic risk. *Frontiers in Artificial Intelligence, 2*, Available: https://doi.org/10.3389/frai.2019.00007.
23. Panneerselvam, S. (2019).Survey of artificial intelligence used in industries a review. *Industrial Engineering Journal, 12*(5), Available:https://doi.org/10.26488/iej.12.5.1188.
24. Park,S.,&Moo,Y. S. (2010).Updating route information in mobile ad hoc networks. *Journal of Mobile Communication, 4*(1), 27–32, Available: https://doi.org/10.3923/jmcomm.2010.27.32.
25. Suganya, M.,&Anandakumar, H. (December 2013).Handover based spectrum allocation in cognitive radio networks. In 2013 International Conference on Green Computing, Communication and Conservation of Energy (ICGCE). doi:https://doi.org/10.1109/icgce.2013.6823431. doi:https://doi.org/10.4018/978-1-5225-5246-8.ch012
26. Rao, M. (1992).Frontiers and challenges of intelligent process control. *Engineering Applications of Artificial Intelligence, 5*(6), 475–481. Available: https://doi.org/10.1016/0952-1976(92)90024-e.
27. Tünay, O. (2010).Chemical oxidation applications for industrial wastewaters. *Water Intelligence Online, 9*, Available: https://doi.org/10.2166/9781780401416.
28. Varaprasad, G. (2012).Stable routing algorithm for mobile ad hoc networks using mobile agent. *International Journal of Communication Systems, 27*(1), 163–170, Available: https://doi.org/10.1002/dac.2354.
29. Võhandu, L. (1994).Artificial intelligence frontiers in statistics. *Engineering Applications of Artificial Intelligence, 7*(1), 87, Available: https://doi.org/10.1016/0952-1976(94)90049-3.
30. Wu, Y.,Deng,S.,&Huang, H. (2012).Information propagation through opportunistic communication in mobile social networks. *Mobile Networks and Applications, 17*(6), 773–781, Available: https://doi.org/10.1007/s11036-012-0401-3.

Chapter 3
Enterprise IoT Modeling: Supervised, Unsupervised, and Reinforcement Learning

Rajesh Kumar Dhanaraj, K. Rajkumar, and U. Hariharan

3.1 Introduction

This chapter introduces the Internet of Things (IoT) and machine learning (ML), including algorithms [1]. According to Statista [26], the worldwide artificial intelligence marketplace totaled $9.51 billion (U.S. dollars) in 2018 and was predicted to reach $14.69 billion in 2019 (Fig. 3.1). Statista also estimated that the number of connected devices in 2019 would total 26.66 billion (Fig. 3.2).

Currently, IoT and ML are fast-growing digital systems. Contemporary systems often intersect, as progress in any kind of technology is no longer achievable in isolation. IoT and ML systems have robust intersections that allow for a variety of strategies, as discussed in the following sections.

3.1.1 Internet of Things

In recent years, the Internet has been extended to nearly all connected products (called "Things") and their virtual representations. The Internet of Things (IoT) has provides for many possible uses, services, and products in numerous areas, including smart homes, healthcare, and smart transportation [2, 3]. Research in this area has received plenty of interest, and naturally plenty of funding. It is supported through the efforts of academia and business, along with standardized systems for a

R. K. Dhanaraj (✉)
Galgotias University, Greater Noida, India

K. Rajkumar · U. Hariharan
Galgotias College of Engineering and Technology, Greater Noida, Uttar Pradesh, India

© Springer Nature Switzerland AG 2020   55
A. Haldorai et al. (eds.), *Business Intelligence for Enterprise Internet of Things*,
EAI/Springer Innovations in Communication and Computing,
https://doi.org/10.1007/978-3-030-44407-5_3

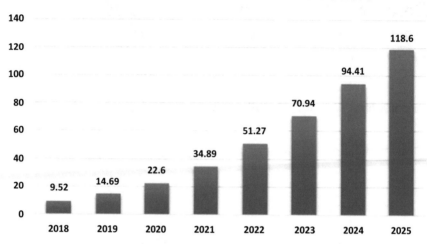

Fig. 3.1 Growth of the artificial intelligence industry

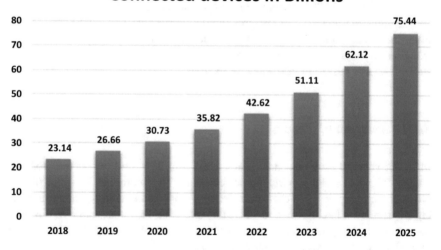

Fig. 3.2 Estimated number of IoT-connected devices worldwide from 2015 to 2025

number of towns in telecommunication, health insurers, the semantic Web, and informatics.

For many years, typical methods were created only for specific uses with minimal versatility. Therefore, when a device was working, it could not be transformed flexibly and dynamically. The first step in introducing the IoT (or more generally, the potential future of the Internet) calls for service platforms, products, and applications that can record, communicate, shop, share, and gain access to information through the actual physical planet; in fact, they are able to speak together around the world. This will likely result in brand-new possibilities for many different domains, including health and fitness, eco-friendly power, manufacturing, smart residences, and personalized end-user apps.

In this way, IoT plays a much more crucial role in daily life. The volume of information on the Web and the Internet is already too much to handle and continues to grow at an incredible pace: Each day, approximately 2.5 quintillion bytes of information are produced. In addition, 90% of the current information was produced in the last few years [4]. Sensory specifics, the info originating from receptors, may be analyzed by means of algorithms and switched into straight in machine info that designs possess a definite understanding about the legitimate data. By doing this, the device is able to act human in some way (some call this type of technique "artificial intelligence"). In addition, we are able to innovate much more useful programs, services, and products that alter our lives dramatically and automatically. For example, meter readings can be used to predict and balance energy usage within smart grids; examining a combination of traffic, pollution, water, and congestion sensory details can provide much better information to site visitors as well as community management; processing and checking sensory products linked to individuals can improve remote health care [5]. Such an information transformation procedure is better illustrated using a "knowledge hierarchy". We have adapted definitions for the concepts of serotonin and semantics (see Fig. 3.1).

3.1.2 Machine Learning

Machine studying (ML) has revolutionized how we do business. A disruptive cutting edge technology which differentiates ML out of various other strategies to hands free operation is actually a level far from the rules-based programming. ML algorithms have made it possible for engineers to use information without having to explicitly program models to examine specific parts of a problem. Rather, the devices themselves return the proper responses, depending on the information they have received [5]. This ability has caused manufacturers to reconsider the methods they normally use to generate choices from the information.

Machine learning is used to generate predictions from new information by using historical details as an instructive case in point. For example, you might wish to determine a consumer lifetime valuation within an eCommerce retailer computing website for an upcoming client meeting. If you currently have historical details on customer interactions and net income related to the buyers, you might wish to make use of machine learning. ML can determine the buyers who are prone to have the greatest net benefit, helping you to concentrate your efforts on them.

Among the strategies used to provide algorithms with information, the most popular design is known as supervised learning. This chapter discusses this area of information science and the reasons why it is regarded as low-hanging fruit for companies that intend to venture into ML.

3.1.3 Machine Learning Plus IoT

ML is being increasing explored by manufacturers because of the hoopla surrounding IoT. Many companies have recognized it as an important strategic area, whereas others are launching pilot programs to determine the possibilities for IoT in their business activities. Consequently, virtually every IT seller is suddenly announcing IoT and consultation services expertise. However, increasing profits by means of IoT is not simple. First, a possible lack of concrete goals is quite disconcerting. Improvements in digitization and IoT locations are brand-new prerequisites with two customers as well as sellers. Many commercial enterprises have neglected to determine how places are going to change together from the setup of an IoT technique.

Quite simply, well-defined, concrete intermediary goals are often absent. For example, manufacturing businesses generate a tremendous quantity of information every day. Nevertheless, businesses do not systematically gather and analyze these data to enhance a procedure's effectiveness or even create additional objectives. In addition, very few vendors actually know how to explain, using concrete terminology, to a prospect the best way to implement beneficial IoT strategies. Just a promise associated with a cloud-based IoT wedge is not sufficient [6, 7].

According to Gartner [27], companies in Finland have advanced to a chapter in which discussions on IoT involve specialized terminology rather than internet business objectives. Consumers are offered revolutionary proposals and are also empowered to take control of the project. However, vendors have to boost their abilities to describe, using a lot more concrete terminology, how businesses can leverage the use of IoT, as well as be inclined to help companies recognize the potential and build practical blueprints. When a seller provides a response on an individual analysis that is too general, alarm bells ought to be ringing.

3.1.3.1 Finding Solutions in IoT Data

A business's brand-new facility was encountering serious production issues resulting from a crucial turbine disaster. A third party was employed to resolve the situation. After approximately 6 weeks, the team of four specialists working with just a field evaluation had made very little progress. The facility required a lot of equipment maintenance and time-stamped environmental data. Even with all of this specific, readily available information, no one was able to transform it into usable data [8, 9].

We assisted the company with an analysis. When all the information was enhanced through analytics, underlying issues were discovered, primarily with much-needed oxygen feeds throughout the manufacturing process. Since this discovery, the facility did not have any major problems. This is a key example of exactly how machine learning can be used to attain much greater effectiveness. With the correct algorithms, a device can be gradually trained to perform internal and external production-related elements, improve the use of consumables, and enhance the effectiveness of the whole output procedure.

3.1.3.2 Machine Learning and IoT in a Business Domain Model

The whole planet is going crazy with information, both artificial intelligence and IoT. Many publications have discussed the quantity of information we produce each and every working day, and countless statistics have demonstrated just how much detail we will produce in the coming years. Thus, we want to discuss how ideas or algorithms coming from other systems can be put into IoT information for optimization. In our previous publications, we discussed data science algorithms with IoT information; here, we discuss machine learning.

On an extremely basic level, Machine learning techniques minimize the man power using training. ML reads patterns in information as a stand-alone device and makes autonomous choices without requiring a developer to create a brand-new group of codes. When you use Siri on your iPhone, for example, you may notice that its replies become more sophisticated and accurate as you use it more often. That is among the fundamental uses of machine learning [10–12].

But exactly how would machine learning improve the IoT? Each time IoT receptors collect information, someone has to be working on the backend to classify the information, process it, and make sure the information is delivered back to the unit for choice generation. When the information generated is huge, exactly how can an analyst take care of the influx? Driverless automobiles, as an example, need to make fast choices when on autopilot, so depending on people is not possible. That is where machine learning becomes useful. To determine which algorithm needs to be used for a specific group of projects, first need determine the job. Several of the activities involve discovering uncommon details, system findings, predicting values and groups, function extraction, and others [7, 13].

Classifying the information sets for various jobs make it much easier for a novice to recognize the proper algorithm program [6]. For example, to focus on information system findings, clustering algorithms, including K means, can be used. K-means was created to deal with substantial chunks of information that include several detail types. In another example, one-class support vector machines and Principal Component Analysis (PCA)-based anomaly detection algorithms are ideal for instruction information coming from uncommon details.

Using IoT and machine learning in the same sentence is like playing buzzword bingo. However, the two principles make good sense as a pair. By 2025, it is believed that the IoT will produce more than 180 zettabytes of information yearly—that is 180 trillion gigabytes. Machine learning is best when used with sizable datasets. To understand why the IoT requires ML to dominate the planet, we begin by examing the two principles individually in Fig. 3.3. Then, we analyze four unique ML and IoT real-life applications. The IoT is developing at an unprecedented speed. In fact, there are 127 brand's -new gadget are connected to the web everyday. Improved connectivity is only going to speed up the progression. McKinsey expects that, by 2022, 100% of the worldwide public will have access to low-power wide area networks (LPWANs) [3, 14, 15]. This can allow long-range marketing communications with attached products while optimizing each expense as well as power consumption demands.

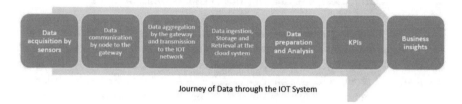

Fig. 3.3 Data generation through an IoT system

3.2 Machine Learning Algorithms

Machine learning algorithms have specific and unique uses. The following list of machine learning algorithms includes AI and developing machine learning systems [8]:

1. Linear Regression
2. Logistic Regression
3. Support Vector Machines
4. Random Forest
5. Naïve Bayes Classification
6. Ordinary Least Square Regression
7. K-means
8. Ensemble Methods
9. Apriori Algorithm
10. Principal Component Analysis
11. Singular Value Decomposition
12. Reinforcement or Semi-Supervised Machine Learning
13. Independent Component Analysis

An understanding of the algorithms will allow you to use them effectively in almost any data problem and a variety of functional machine learning projects.

The following three categories apply to machine learning algorithms [16]:

1. Supervised Learning
2. Unsupervised Learning
3. Reinforcement Learning

It is important to understand each algorithm in order to select the correct one to meet your problem and learning requirements, as shown in Fig. 3.4.

3.2.1 Supervised Learning

Supervised learning techniques are used by optimum machine learning users. In supervised learning, an algorithm's learning process is completed with an instruction dataset. Even by making use of a training dataset, the task could be regarded as

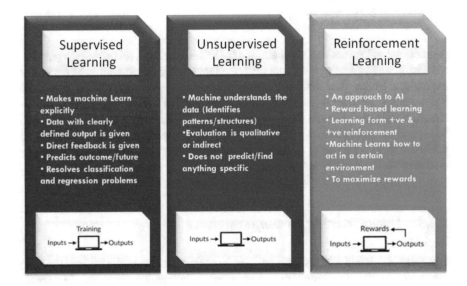

Fig. 3.4 Types of machine learning

a mentor that is supervising the learning process. The algorithm can help within generating predictions regarding the data in the training process and obtain the modifications carried out by the teacher itself. A conclusion may indicate if the algorithm has accomplished an appropriate amount or a certain degree of performance. There are two kinds of supervised learning problems, which can be more grouped generally as regression or classification issues:

- **Classification Problems:** A problem which generates adaptable, which moves lower merely especially such as the "red" or perhaps "blue" or perhaps it may be "disease" and "no disease".
- **Regression Problems:** A regression concern is whether the variables are really authentic, such as "dollars" or perhaps it might be "weight" [16, 17].

Several issues can be noted within the data type, including things like time series prediction and recommendations. Supervised machine learning algorithms including the following:

1. Decision Trees
2. Naive Bayes Classification
3. Support vector machines for classification problems
4. Random forest for classification and regression problems
5. Linear regression for regression problems
6. Ordinary Least Squares Regression
7. Logistic Regression
8. Ensemble Methods

3.2.2 Decision Trees

A great deal of information is apparent in the name decision tree; using easy phrases, a decision tree can help you make a choice regarding the information at hand. For example, a banker can make a decision on providing a mortgage to an individual based on age, profession, and amount using a decision tree. When creating a decision tree, the task is launched in root node, responds to certain questions within each node, and considers the area that corresponds to the specific solution. Adhering to the procedure, we traverse to the root node, then to a leaf, then type conclusions within the context of the information. An example is provided in Fig. 3.5.

3.2.3 Unsupervised Learning

Unsupervised learning uses an algorithm in which you insert the information (X) along with hardly, no any corresponding variables are getting established. The main objective for unsupervised learning is to help model the underlying framework or in an effort to assist the learners in understanding the information. Unsupervised learning is different from supervised learning because there are no any proper responses are not mentioned the numbers to the instructor. Algorithms remain as their very own products to help you find the intriguing framework that is contained in the information. Unsupervised learning may be grouped as clustering or connection issues [7]:

Fig. 3.5 A tree showing the survival of passengers on the Titanic

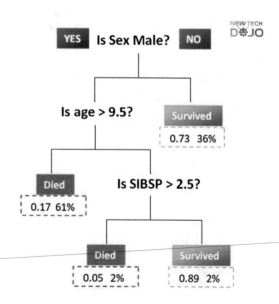

1. **Clustering:** A clustering shows what you would like to find out and also assists in determining the natural groups belonging to the information, such grouping clients according to their buying behavior.
2. **Association:** A connection becomes the learning issue. It is exactly where you will find the guidelines that will explain the larger areas of your information. For example, individuals who purchase X are also individuals who have a tendency to purchase Y.

Some common unsupervised learning algorithms include the following:

- K means for clustering problems
- Apriori algorithm for connection guideline learning problems
- Principal Component Analysis
- Singular Value Decomposition
- Independent Component Analysis

3.2.4 Reinforcement or Semi-supervised Machine Learning

For some problems, you will need to enter a huge amount of information. In semi-supervised machine learning, the information is first labelled as (X), then later some is labelled as (Y). This method is somewhere between supervised learning and unsupervised learning. A great example is an image archive in which the subjects of only some photos are labelled (e.g., dog, cat, most people, but a school is unlabelled) [4].

A great deal of practical society items learning related problems falls straight into this specific group because they might be costly or time consuming. To label the information, access may need to be obtained through the domain name professionals. The unlabeled information is low-cost also comparatively simple to gather as well as find. Unsupervised learning strategies can be used to accomplish great things. They can help you learn as well as find out the different legitimate components that occur in the type of variable. Supervised learning methods help make the very best of estimate predictions with unlabeled information. This information can then be used in supervised learning algorithms as instructive information to make predictions based on different details, as shown in Fig. 3.6.

3.3 Machine Learning Applications in IoT

3.3.1 Price Savings Come to Industrial Applications

Predictive abilities are incredibly beneficial in a manufacturing environment. By taking in information from several sources within and on devices, machine learning algorithms can "learn" what is common for any device then identify when something abnormal starts to happen. "The collected information is delivered to the serv-

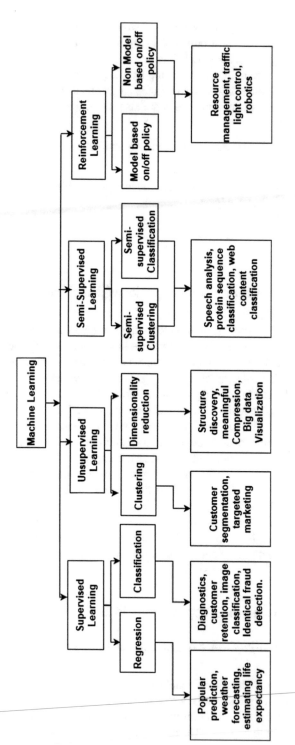

Fig. 3.6 Overall classification of machine learning

ers, exactly where it's contrast to earlier details collected out of this printer, and even information collected by using the same devices very much. The wedge of ours is able to identify the smallest alterations as well as alert you of acquiring malfunctions. This particular evaluation is actually carried out within real time and also the outcomes are actually shown on the technician's smart phone in just seconds." [7]

Predicting if a machine requires upkeep can be quite important, converting directly into money by saved expenses. An excellent case is Goldcorp, a mining business that uses surprisingly impressive cars are dragged to the bay. When these transporting automobiles decompose, Goldcorp is losing 2 billion dollars each day from inefficiency [13]. Goldcorp is currently making use of machine learning to predict with more than 90% reliability when devices will require upkeep, which results in a large cost savings.

3.3.2 Shaping Experiences to Individuals

Almost everyone is familiar with machine learning in their daily lives. Netflix and Amazon use machine learning to understand our tastes and provide a much better experience for their end users. Thus, they suggest products that you might want or provide suggestions for films and television shows to watch.

Likewise, IoT machine learning can be very important in shaping our environments to our personal preferences. The Nest Thermostat is a terrific illustration, as it uses machine learning to study your tastes for cooling and heating, ensuring that the home is the perfect temperature whenever you return from work or awaken each morning.

3.3.3 The New Innovative IoT Emerging Business Models

Like each main technological shift, the growth of the IoT has been both successful and disastrous. The passion with which business owners, investors, and governments have adopted IoT is a testament to its deep cultural and economic possibilities.

Nonetheless, passion has outpaced comprehension. Consequently, the IoT business is rife with all sorts of errors and misapplications that typically plague early adoption. As IoT moves beyond its infancy stage, the economic realities are becoming more clear. Probably the most considerable of the revelations is the fact that conventional hardware does not conveniently affect IoT systems [3, 15] because IoT products offer recurring, infrastructure-related expenses unlike conventional hardware. Consequently, IoT systems are only confirmed to be beneficial when the software leads to recurring, constant value for a client. This value is usually accomplished by a number of methods that are distinct, including elevated advantages, decreased operating expenses, and facilitation of compliance. With this section, we examine the

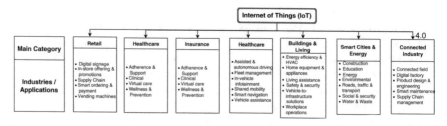

Fig. 3.7 Applications of the IoT

special economics of IoT and its most promising industrial uses. We also incorporate particular use cases to help you understand the applications, as shown in Fig. 3.7.

3.3.4 The Economics of IoT Using Machine Learning

The standard hardware industry is based upon the traditional, product-at-a-margin principal. A business creates a portion of hardware for X and also sells it to customers for X + Y. What complicates the unit inside IoT is the fact that the X value is usually snowballing as well as recurring. That is, IoT items incur constant infrastructure-related expenses that do not affect conventional hardware. Therefore, to stay operational, IoT products need a network where to run, along with a system wedge where to gather as well as control information. Consequently, conventional device economics easily fail on a company's scale or as equipment ages. IoT businesses are actually switching to business to shield their margins from erosion. These designs frequently depend on recurring payments near the buyer, such as to the "hardware as a service" industry version. Hayesis placed a unit whereby a hardware company leased the item to clients for a monthly or annual fee. IoT wedge suppliers, meanwhile, charge monthly fees because of the compilation, business, and creation of device-driven details. Due to these recurring expenses, IoT systems are economically beneficial whenever they provide recurring, constant worth to a client. Thus, there may be a typical group of problems preventing a connected unit from becoming successful:

- Model has an annuity (recurring earnings stream)
- Consistent or even improved sales through association with a consumable
- Direct and indirect operational effectiveness gains
- Exponential progress of perpetuity

A product has an incredibly scant, non-variable life cycle. Opening of various earnings by the way of info put together. In other circumstances, an online business offering linked treatments with an ordinary design will probably be unsuccessful. That is because, with time, the recurring backend expenses will erode with the business's margins. Within these sections, we examine the most promising industrial uses of IoT systems, as well as describe the apps with distinct use case examples [18].

3.3.5 Compliance Monitoring

Every year, American manufacturers invest an estimated $192 billion to comply with economic, environmental, and work safety regulations. IoT systems have shown strong possibilities in this realm. By remotely overseeing very sensitive assets, IoT products can enable companies to considerably reduce expenses related to regulatory compliance. A good example is software used in the oil and gas industry. Gas and oil extraction and processing are governed by strict compliance requirements. Environmental and safety laws require constant vigilance to uphold. However, internet-connected equipment makes compliance considerably less complicated and cheaper. In the past, an area agent would have to actually examine an extraction or processing site to confirm regulatory compliance. Today, an IoT unit can be used in the area to remotely check vital compliance metrics, such as oil leaks and gasoline pollutants. Not only does this decrease the expenses for onsite tracking, but it is also a lot more responsive. Although an area agent might be in a position to go to a certain site occasionally, an IoT unit is able to offer regular, up-to-the-minute details in real time. This responsiveness also helps to limit the scope of penalties when noncompliance occurs. For pollutants and leakages, regulatory penalties are usually proportional to the quantity of materials produced. IoT products might limit a business's risk for penalties by allowing to to react faster to leaks.

IoT systems also are proving to be helpful with facilitating compliance within the insurance industry. Insurance companies are using IoT products to monitor elements of policy adherence. A particularly intriguing illustration of this use can be found in the homeowner's insurance sector, where insurers are using IoT products to monitor fluids amounts within basements. When water starts to build up in a basement, the IoT unit sends notifications to the insurer and property owner, so that they can correct the problem before substantial deterioration occurs. This allows the insurer to decrease the price of policies and protects homeowners by preventing damage to their homes. For these reasons, IoT systems may decrease the overhead associated with compliance and considerably decrease a business's costs associated with fines or policy payouts, for example [19], as shown in Fig. 3.8.

3.3.6 Preventative Maintenance

The expenses related to equipment malfunction have a tendency to cascade. Just one malfunctioning device can impede production, interrupt a whole supply chain, or, as discussed previously, lead to regulatory noncompliance. Fortunately, the same features that allow IoT systems to facilitate compliance also help companies protect invaluable assets. Returning to the use case of gasoline and petroleum extraction, we can see how IoT systems help to revolutionize equipment upkeep. Petroleum extraction sites are full of expensive and highly complex products. Typically, these

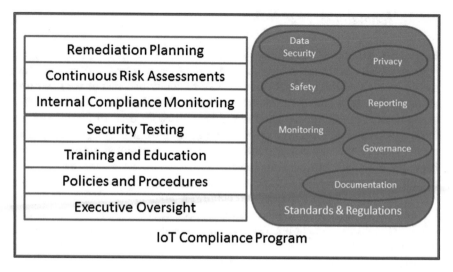

Fig. 3.8 IoT compliance monitoring

products were safeguarded exclusively by occasional and possibly imperfect field inspections. Because of their intricacy and scope, even small technological mal-functions could to lead to decreased production and substantial monetary losses. Catastrophic disasters, meanwhile, could cost companies countless dollars. Today, a low-cost IoT unit can be used to remotely keep track of gear, monitor upkeep schedules, and protect against malfunctions. When a machine part starts to under-perform or maybe malfunction, the IoT unit will immediately issue an automatic alert. Doing this will restrict losses caused by inefficiency, as well as prevent large-scale problems and injuries [9].

3.3.7 Remote Diagnostics

The remote monitoring abilities of IoT technology can be effectively used outside of the area of physical upkeep. IoT products are now being used to obtain compre-hensive and regular analytic details in several areas, ranging from pharmaceuticals to agriculture. The compilation of data is allowing companies to automate some procedures and remotely manage manufacturing. A good illustration can be found in the agricultural sector. Forward-thinking farmers have used IoT products to mon-itor crops. A connected unit can constantly monitor environmental conditions, such as soil conditions, sunlight, humidity, and temperature. This data is then compiled and analyzed to help farmers assess long-term environmental conditions. Additional IoT products may be used to immediately react to this information. For example, when a device registers very low moisture in the soil, it can instantly apply water to fix it. These features increase efficiency and decrease overhead expenses. Just like other IoT programs, the outcome is better, more effective, and less costly [20].

3.3.8 Asset Tracking

Visibility is easily the most useful commodity within supply chain control. However, more than 60% of businesses do not have complete supply chain visibility; the typical inventory precision threshold of U.S. merchants is only 63%. Fortunately, low-cost IoT systems now allow companies to overcome this issue. A basic 3G-connected microcontroller can find and monitor inventory in real time, from anywhere on Earth. This unprecedented level of visibility will help businesses to prevent theft and loss, enhance fleet advantages, and improve forecasting. Monitoring information can be shared immediately with other entities in a supply chain to additionally boost visibility and decrease inefficiencies. Cisco and DHL, for example, expect that IoT systems can have a $1.9 trillion effect on their strategies and the supply chain management industry. To help illustrate the possibilities, Fig. 3.9 shows IoT-based advantage monitoring that benefits an simple, linear supply chain.

3.3.9 Automatic Fulfillment

Improved visibility on the customer end of this resource chain also allows small businesses to carry out brand new, much more profitable programs. Up-to-the-minute inventory as well as usage information produces automated satisfaction and price agreements for new buyers. This performance can help companies by improving need forecasting, generating product sales, and informing marketing and advertising efforts. In doing so, customers can benefit by experiencing less merchandise shortages and more effective customer support. A fascinating example of this IoT program is the Amazon Dash button, which is a small IoT unit that can be set up to reorder certain products when pressed. For instance, a consumer can easily use their button to

Manufacturer	Distributor	Retailer	Customer
Accurate, up-to-date inventory monitoring to inform production & eliminate stock outages	In-transit tracking allows distributors to anticipate & avoid delays	Anticipate & accommodate upstream delays	Gain insight into customer usage & behaviors
Real-time raw materials monitoring to prevent production downtime	Collection and analysis of in-transit data allows for calculation of more efficient fleet routes	Monitor in-house inventory to prevent stock outages or surpluses	Monitor product status to ensure quality & integrity
Improved warehouse management & organization by reducing lost, misplaced, & inefficiently-spaced assets	Real-time pickup & delivery notifications	Real-time sales and returns data to improve demand forecasting	Provide live tracking & delivery notifications direct to customer
	Round-the-clock asset tracking leads to reduced risk of loss and theft	Monitor floor stock to ensure product visibility	

Fig. 3.9 Asset tracking

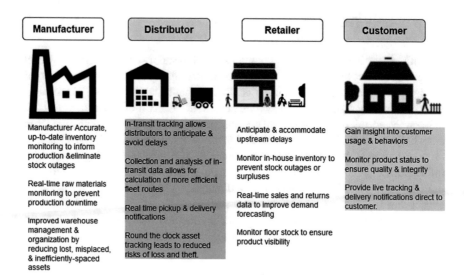

Fig. 3.10 Example of automated fulfillment

purchase a 12-pack of Bounty paper towels anytime it is pressed. Each switch is pro-grammed for just one item and amount at any given time. The buttons can be mounted with a user's home, near each item's location. Not only do the designs help to improve product sales, but they supply priceless details on client consumption practices. These system designs are still emerging, but their possibilities are clear. At the end of 2016, Amazon announced that Dash gross sales had increased by more than 400% com-pared with 2015, with 60 new brands signing on to participate in the program. Even though the Dash design depends on client feedback to trigger replenishment, genu-inely automated devices have started to gain traction, as shown in Fig. 3.10.

3.3.10 Adding Value to the IoT with AI and ML

Artificial intelligence may be used hand in hand with a different technology, machine learning. Often used interchangeably, the terms AI and ML represent the basic principle of acquiring applications that have intelligence. This particular intel-ligence enables them to evaluate information and generate choices much like the human mind. The essence of IoT products is to gather information and generate uses from it; information from physical products via AI and ML allows us to expand these procedures. The Internet of Intelligent Things (IoIT) uses artificial intelli-gence takes importance over the IoT domain for higher interpreting information about products that are connected [13].

The products within an IoT network are connected through receptors, actuators, hardware, and software programs to provide people with rational inputs. The basis of the IoT is AI and ML because it enables the products to interpret the information collected. Whenever connected devices gather and compile raw details, the applications empowered with ML abilities can merge and assess the information [20].

3.3.11 Advantages of Using IoT and ML Together

The IoIT produces IoT programs that recognize their potential. AI and ML allow for much more comprehensive insight at a quicker speed. Businesses are implementing IoIT to the benefits discussed in the following sections.

3.3.12 Improved Accuracy Rate

If you have ever attempted to evaluate information in several different spreadsheets, you know it can be a tiresome task. Human brains are restricted to performing particular duties at a particular speed, so when our brains are tired, we are likely to make mistakes. The IoT has the capacity to process large amounts of information and create an analysis. The entire process is machine and software driven; it can be performed with no human intervention, which tends to make it error free and also increases reliability [26].

E-commerce transactions, online payments, and ATM withdrawals, for example, are very susceptible to fraud. Using the consolidated strength of human comprehension and IoT machine learning as well as Robotic Process Automation (RPA) methods of manmade intelligence, fraud may be prevented, therefore stopping some monetary losses.

3.3.13 Predictive Analysis and Maintenance

Predictive analytics is a type of evaluation that examines pre-existing information and predicts probable succeeding outcomes. It would not be an exaggeration to say that IoT and AI are the basis of predictive maintenance. Presently, IoT products are being used by businesses to monitors for malfunctions or concerns in an automatic way with no human intervention. By using AI, this process can allow models to perform a predictive evaluation, giving businesses the ability to identify possible mishaps, malfunctions, and advance maintenance. For this reason, the risk of losses is extremely reduced because problems are now being recognized in advance.

For example, delivery businesses can make use of predictive evaluation to examine and analyze their information to avoid unexpected downtime for shipping and maintain shipping by way of frequent servicing [20].

3.3.14 Improved Customer Satisfaction

The center of each company is client satisfaction. Presently, organizations such as Amazon have gained recognition for being the best customer-centric businesses by keeping the goals of their clients above all else. Nevertheless, human-based consumer encounter may not be successful for a number of reasons, including language barriers and time restrictions [20].

Businesses are realizing the advantages of AI by using chatbots to interact with clients. Great quantities of client information can be used to supply them with a far more personalized experience, as well as answering their queries appropriately.

3.3.15 Increased Operational Efficiency

Predictions produced via manmade intelligence are extremely beneficial for increasing the operational effectiveness of a company. In-depth insights received via manmade intelligence are often used to enhance a company's processes, which may lead to enhanced operational effectiveness plus reduced expenses.

With exact predictions, insights can be obtained concerning price consuming matters as well as the period of the housing industry also automates these to enhance the effective numbers. Additionally, for shipping businesses, the insights received via man-made intelligence can help to improve procedures, equipment options, and inventory to reduce needless expenditures [25].

- **Improve the basic safety of equipment** – The IoT can keep track of all machinery, from compressors to high-speed engines, to quickly recognized potential problems. This can result in greater safety for staff members and even the environment.
- **Decrease lost revenue related to downtime** – Downtime is costly. By using ML and the IoT, problems can be recognized in advance to considerably decrease downtime, thus minimizing lost revenue.
- **Allow for correct record keeping** – The IIoT makes use of the information to keep track of equipment, flag questionable components, and determine what maintenance is needed.
- **Predict equipment failure before it occurs** – Real-time data allows a company to assess what is occurring, but the IIoT is able to take action on your behalf. Using man-made intelligence, algorithms, and much more, information analyzers are now able to predict and alert you to potential failures.
- **Provide real-time data** – With ML analytics, a company can obtain a data evaluation in real time to ensure that things are operating properly and that the business is flourishing.

3.4   Challenges in IoT Implementation

Problems in IoT implementation are related to network security expenses and information evaluation. Most businesses may be ready to deal with sensors and data analytics, but they neglect to focus the same attention on network investments, system integration, and security. This gap highlights a rift between technologies in addition to excellent data transfer, where IT division is really sharp on buying the technical feature needed having a prosperous IoT undertaking, it is the lack of really worth sent, which is likely to create business experts look at the unsuccessful endeavor. Capturing invaluable info that leads to phony attention could also work as the primary reason a great deal of business experts have a look at their IoT projects unsuccessful. Problems experienced by businesses that create quality problems, resulting in the venture disaster, which includes night conclusion as well as finances overruns occasions [20].

3.4.1   Interoperability and Compatibility of Various IoT Systems

According to the marketplace analysts with McKinsey, 40–60% of overall value is based on the ability to attain interoperability between various IoT methods. With many vendors, service providers, and OEMs, it becomes very difficult to preserve interoperability between various IoT methods. Receptors and networking are the essential parts of the IoT. However, not every single machine comes with complex receptors and social networking abilities. Also, receptors with different energy consumption and security requirements, built within history devices, might not be able to offer the exact same outcomes.

A fast workaround might be to add external receptors. However, this is also difficult because it must be determined what components and data will be communicated to the network.

3.4.2   Authentication and Identification of Technologies

Currently, there are approximately 20 billion linked products in existence, as well as in order to feel foundation of all the gear requires a good offer of safeguarding implications without having just intricacy. Using many devices that are connected on a single wedge requires formalization or a method to authenticate the gadgets [26].

3.4.3   Integration of IoT Products with IoT Platforms

For a profitable setup of an IoT program, businesses have to incorporate a variety of IoT connected goods with correct IoT platforms. Inappropriate integration could result in irregularities in productivity along with executing to move really worth towards the customers. A vice president with Gartner, Benoit Lheureux [27], claimed that "Through 2018, 50 % of the expense of applying IoT remedies will probably be invested combining different IoT parts with every back-end and other methods. It's essential to be aware of integration is actually an important IoT competency."

However, there may be too many IoT endpoints to aggregate the sensor information or transmit it to an IoT wedge. With rich integration, businesses are able to mine the massive information by way of Big Data methods in order to produce awareness and predict results [19].

3.4.4   Connectivity

Connectivity is a common marketing difficulty because the Internet is not available everywhere with the same speed. A worldwide satellite business, Inmarsat, disclosed that 24% of users reported connectivity problems as one of the greatest problems within an IoT deployment. Strong networks are needed to gather information, but they may experience difficulties in transmitting data.

The caliber of indicators collected by the receptors and transmitted to the networks mostly depend on the routers—MAN, LAN, and also WAN. These networks have to be well connected via various systems to facilitate fast and high-quality reception. Though the variety of gadgets that are connected to the networking coverage, and that generates keeping track of as well as keeping track of issues.

3.4.5   Handling Unstructured Data

An increasing number of connected products also increase the difficulties of dealing with unstructured details on the volume, velocity, and type. However, an actual struggle for businesses is determining what information is valuable, as only quality information is actionable. Unstructured details cannot be kept in an SQL structure. Unstructured information kept in a NoSQL structure makes the retrieval of information a little complicated. With the launch of Big Data frameworks, such as Cassandra and Hadoop, the problems with unstructured details have decreased. However, merging the fundamental data with IoT causes problems. In addition, at this time, there are not any common recommendations for the use and retention of information or metadata.

3.4.6 Data Capturing Capabilities

A challenge of recording information is to transform the information collected from different sources into a common structure that can be analyzed and automated. According to a HubSpot article, sponsored by Par Stream [25], 86% of company stakeholders assert that information is essential to their IoT tasks, but just 8% are actually able to shoot as well as evaluate IoTin formation wearing a regular fashion, as shown in Fig. 3.11.

The IoT depends on receptors for networks and indicators for that division, so particular anomalies in deep runtime, like a power outage, could cause incorrect data to be captured.

3.4.7 Intelligent Analytics

The intent behind IoT is to convert data into substantial information. Flaws in the data or information design could result in false conclusions. We have to understand the info inside itself is not a comprehension, type of right problems have to become guided from the actual info reach the understanding.

According to an article by Hubspot [25], 42% of IoT stakeholders experience issues with recorded information, and 30% stated that their analytic abilities were not effective or adaptable, as shown in Fig. 3.12.

Historical methods, including standard analytics programs in which only limited information can be examined at any given time, can certainly restrict the ability to control real-time details. The following are some difficulties that prevent smart analytics:

- Unpredictable activity of a computer throughout an incident.
- Traditional analytics software.

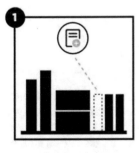

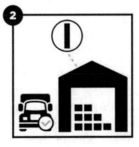

Sensor detects low quantity of product X. Sends alert to cloud.

Cloud relays alert message to distribution for replacement product.

Consumer receives product before running out. Merchant profits from automatic recurring sale.

Fig. 3.11 Data capturing

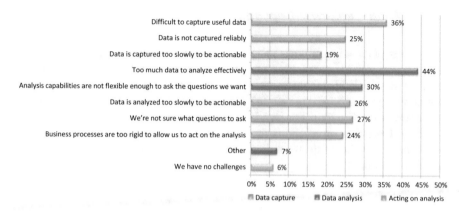

Fig. 3.12 Challenges with collecting and analyzing data in an IoT project

- Slow adoption of current engineering as a result of the excessive cost.
- Lack of competent workers in information mining, algorithms, machine learning, and complicated processing [26].

3.4.8 Data Security and Privacy Issues

Even prominent companies such as Apple and visionaries such as Elon Musk have not been spared by online hackers. Recent occurrences of ransomware strikes have inhibited the confidence of many companies. One analysis reported that by 2020, 25% of cyber-attacks will focus on IoT products [24]:

- Malware infiltration: 24%
- Phishing attacks: 24%
- Social engineering attacks: 18%
- Device misconfiguration issues: 11%
- Privilege escalation: 9%
- Credential theft: 6%

Lapses in cyber-safety could occur in any organization and affect customers. Therefore, it is crucial for every business to take measures to enhance security. A report disclosed that 45% of IoT unit proprietors do not use any kind of third-party protection application. In addition, 35% of these individuals do not change the default password on their gadgets. Thus, buyers and businesses must collaborate to implement effective security policies in IoT setup [20].

3.4.9   Consumer Awareness

Many individuals are not aware of the IoT, although they fully understand the dependency on smart apps for information, stocks, and entertainment. It is not really essential for customers to understand how elements function commercially. However, a lack of fundamental knowledge may cause anxiety about price and security, which may result in the slow adoption of technologies.

As outlined in a survey of 3000 Canadian and U.S. customers carried out by Cisco, 53% of customers would not choose to individual information collected, notwithstanding of this unit. This indicates that subscribers are reluctant to disclose their information, which may serve as a deterrent to IoT implementation [20].

3.4.10   Delivering Value

Based on a Forbes Insights Survey[23], 29% of professionals are feeling a significant struggle in creating IoT abilities stands out as the quality of IoT. This information indicates the challenge of IoT program growth businesses with obtaining valuation for their customers. Thus, prior to plunging straight into the improvement of IoT programs, a business needs to determine what value they are likely to provide via what abilities, as well as how the solution will enhance effectiveness and efficiency, while simultaneously producing customer satisfaction.

Because the IoT is about connected things, the IoT tasks also need a great degree of help in this manner. Approximately 50% of businesses with IoT initiatives are clearly associated with IT services, suppliers, or consultation services companies, depending upon them to assist with shipping, delivery, and supply company recommendations. Feel foundation, creating an IoT Development Company, that thinks engineering away from labor as well as like function by merging all the components of IoT with in a manner that is based on connectivity, improving recognition and holding accuracy of all of the phases [20]. Nevertheless, continue a range of enhancements of produce abilities to enhance the effectiveness of the item or maybe system depending on the most recent engineering.

3.5   Conclusion

The modern era of the IoT and AI will greatly improve existing tasks. With hands-free operation and in-depth evaluations, companies can enjoy the advantages of progress while maximizing profits. The challenges of this era are to make better use of AI and the IoT for the new generation. The distinctive, infrastructure-related expenses of IoT systems is required for predicting the consistent hardware models.

With these brand-new versions, IoT companies need to provide recurring value to their clients in order to achieve success. Effective IoT businesses are delivering recurring value to their customers by applying service-based internet business versions and monetizing consumer information. To ensure that the designs catch the attention of clients, successful IoT companies should also supply recurring value in exchange. The value can be made available in a number of ways. As the IoT business matures, distinct products have started to appear. IoT products can supply recurring value by improving operational efficiencies, facilitating compliance, enhancing sales, and opening new sales channels.

References

1. Kumar, D. R., Krishna, T. A., & Wahi, A. (2018). Health monitoring framework for in time recognition of pulmonary embolism using internet of things. *Journal of Computational and Theoretical Nanoscience, 15*(5), 1598–1602.
2. Anandhalli, M., & Baligar, V. P. (2017). A novel approach in real-time vehicle detection and tracking using raspberry pi. *Alexandria Engineering Journal, 57*(3), 1597–1607.
3. Atzori, L., Iera, A., & Morabito, G. (2010). The internet of things: A survey. *Computer Networks, 54*(15), 2787–2805.
4. Kamath, G., Agnihotri, P., Valero, M., Sarker, K., & Song, W. -Z. (2016). Pushing analytics to the edge. In *GLOBECOM*, IEEE, pp 1–6.
5. Deepak, S., & Anandakumar, H. (2019). AODV route discovery and route maintenance in MANETs. In *2019 5th International Conference on Advanced Computing & Communication Systems (ICACCS)*. https://doi.org/10.1109/icaccs.2019.8728456.
6. Nishiguchi Y, Yano A., Ohtani, T., Matsukura, R., & Kakuta, J. (2018). Iotfault management platform with device virtualization. In *2018 IEEE4th World Forum on Internet of Things (WF-IoT)*, pp. 257–262.
7. Qiu, J., Wu, Q., Ding, G., Xu, Y., & Feng, S. (2016). A survey of machine learning for big data processing. *EURASIP Journal on Advances in Signal Processing, 2016*, 67.
8. Redmon, J., Divvala, S., Girshick, R., & Farhadi, A. (2016). You only look once: Unified, real-time object detection. In *The IEEE conference on Computer Vision and Pattern Recognition (CVPR)*, pp. 779–788.
9. Kumar, R. N., Karthick, S., Valarmathi, R. S., & Kumar, D. R. (2018). Design and analysis of multiply and accumulation units using low power adders. *Journal of Computational and Theoretical Nanoscience, 15*(5), 1712–1718.
10. Anandakumar, H., & Umamaheswari, K. (2017). Supervised machine learning techniques in cognitive radio networks during cooperative spectrum handovers. *Cluster Computing, 20*(2), 1505–1515.
11. Howard, A. G., Zhu, M., Chen, B., Kalenichenko, D., Wang, W., Weyand, T., Andreetto, M., & Adam, H. (2017). *Mobilenets: Efficient convolutional neural networks for mobile vision applications*. Computer Vision and Pattern Recognition (cs.CV).
12. Iandola, F. N., Han, S., Moskewicz, M. W., Ashraf, K., Dally, W. J., & Keutzer, K. (2016). *Squeezenet: Alexnet-level accuracy with 50x fewer parameters and <05mbmodel size.* arXiv:1602.07360
13. Rajesh Kumar, D., & Shanmugam, A. (2017). A hyper heuristic localization based cloned node detection technique using GSA based simulated annealing in sensor networks. In *Cognitive computing for big data systems over IoT*, pp. 307–335.

14. Bansod, G., Raval, N., & Pisharoty, N. (2015). Implementation of a new light weight encryption design for embedded security. *IEEE Transactions on Information Forensics and Security, 10*, 142–151.
15. Cecchinel, C., Jimenez, M., Mosser, S., & Riveill, M. (2014). An architecture to support the collection of big data in the internet of things. In: *2014 IEEE World congress on services*, IEEE, pp. 442–449.
16. Anandakumar, H., & Umamaheswari, K. (2018). A bio-inspired swarm intelligence technique for social aware cognitive radio handovers. *Computers and Electrical Engineering, 71*, 925–937. https://doi.org/10.1016/j.compeleceng.2017.09.016.
17. Varghese, B., Wang, N., Barbhuiya, S., Kilpatrick, P., & Nikolopoulos, D. S. (2016, November 18–20). Challenges and opportunities in edge computing. In *2016 IEEE international conference on smart cloud, Smart Cloud 2016*, New York.
18. Zhang, D., et al. (2018). Review on the research and practice of deep learning and reinforcement learning in smart grids. *CSEE Journal of Power and Energy Systems, 4*, 362–370.
19. Chen, G., Parada, C., & Heigold, G. (2014). Small-footprint keyword spotting using deep neural networks. In *ICASSP*, IEEE, pp 4087–4091.
20. Doshi, R., Apthorpe, N., & Feamster, N. (2018, May). Machine learning DDos detection for consumer internet of things devices. In *2018 IEEE Security and Privacy Workshops (SPW)*, pp. 29–35.
21. Dastjerdi, A. V., & Buyya, R. (2016). Fog computing: Helping the internet of things realize its potential. *IEEE Computer, 49*(8), 112–116.
22. Dhiviya, S., Malathy, S., & Kumar, D. R. (2018). Internet of things (IoT) elements, trends and applications. *Journal of Computational and Theoretical Nanoscience, 15*(5), 1639–1643.
23. https://www.forbes.com/forbes-insights/our-work/internet-of-things/
24. Ioannis Stellios, Panayiotis Kotzanikolaou, Mihalis Psarakis, Cristina Alcaraz, Javier Lopez, A Survey of IoT-Enabled Cyberattacks: Assessing Attack Paths to Critical Infrastructures and Services. IEEE Communications Surveys & Tutorials 20 (4):3453-3495
25. https://blog.hubspot.com/marketing/iot-retail
26. https://www.statista.com/statistics/607716/worldwide-artificial-intelligence-market-revenues/
27. https://www.gartner.com/smarterwithgartner/five-steps-to-address-iot-integration-challenges/

Chapter 4
An Overall Perspective on Establishing End-to-End Security in Enterprise IoT (E-IoT)

Vidya Rao, K. V. Prema, and Shreyas Suresh Rao

4.1 Introduction

IoT is a vast network of networks consisting of physical and virtual interconnected entities. These entities have unique addressing schemes and interact with each other to provide certain customized or generic services. In 2012, the International Telecommunication Union (ITU-T) recommended a standard definition of IoT as "a global infrastructure for the information society, enabling advanced services by interconnecting virtual and physical things based on existing and evolving interoperable information and communication technologies [1]". Technically speaking, IoT has its applications in diverse areas like healthcare, surveillance, transport, security, manufacturing, environmental monitoring, and food processing, and it is integrated with technologies like autonomic networking, decision making, machine-to-machine communication, cloud computing, big data analytics, confidentiality protection, and security [2].

Enterprise Internet of Things (E-IoT) is the next level of sensor technology that connects every physical object to form a vast network of embedded computing devices. These devices are generally made up of tiny components. They have constrained processing capabilities, low memory, and limited power resources. This emerging technology has reduced manual intervention and has increased business efficacy.

Gartner Press released an article in August 2019 showcasing that by 2020, there would be about 5.8 billion IoT endpoints, as compared to 4.8 billion endpoints during 2019. That means there is almost a 21% increase in the addition of new endpoints.

V. Rao (✉) · K. V. Prema
Manipal Institute of Technology, Manipal Academy of Higher Education,
Manipal, Karnataka, India
e-mail: prema.kv@manipal.edu

S. S. Rao
Sahyadri College of Engineering and Management, Mangalore, Karnataka, India

© Springer Nature Switzerland AG 2020
A. Haldorai et al. (eds.), *Business Intelligence for Enterprise Internet of Things*,
EAI/Springer Innovations in Communication and Computing,
https://doi.org/10.1007/978-3-030-44407-5_4

These endpoints are categorized under various use cases like utilities, government buildings, automation, physical security, healthcare providers, manufacturing and natural resources, information and transportation, retail, and wholesale. Among these use cases, utilities have taken a major share of 17% with 1.33 billion endpoints with applications like electric smart grid, smart metering, and smart electricity supply. Apart from this, physical security application surveillances, intruder systems, as well as CCTVs have taken about 0.70 billion endpoints.

These endpoints have generated a total revenue of about $389 billion in countries like North America (NA), Greater China (GC), and Western Europe (WE). Statistics of Gartner's study have shown that about 75% of the revenue would be generated by electronic endpoints in the world, i.e., about $120 billion revenue from NA and $91 billion and $82 billion revenue from GC and WE, respectively, by 2020. It is expected that the two main use cases that shall take a good share in the electronic revenue are connected to consumer cars and networkable printing and photocopying with $71 billion and $38 billion revenue, respectively. Then comes the government indoor and outdoor surveillances that add on to the revenue as the government is considering civilian security as its top priority.

These endpoints are enabled with various sensors like cameras, proximity sensors, temperature sensors, air quality sensors, flow sensors, and many more sensors that are unprotected. The reason is that the agent-based technologies do not protect them from various attacks like distributed denial of service (DDoS), ransomware attack, stealing of sensitive intellectual properties, cryptojacking attacks, etc. [3]. This is a major cause of concern as the data produced by these devices consists of user health information, bank details, passwords, location information, and many more. Hence these devices are subjected to security threats due to (a) malicious or compromised node in the network, (b) defective manufacturing, and (c) presence of an external adversary. There may also be threats to security initiated by nature. These natural threats include earthquakes, floods, fire, and hurricanes that cause severe damage to the computer systems. As it is hard to safeguard against natural calamities, it is advisable to reduce the damage by collecting backup of data through a contingency plan. Similarly, there could be human threats that can be classified under information-level attacks, adversary location attacks, access-level attacks, and host-based attacks. To enable the security of the devices, it is essential to select the hardware components that have the following properties: default authentication capabilities, end-to-end traffic encryption, secure boot loading process, enforcement of digital signatures during firmware update, and transparent transactions.

Also, it has been identified that there are almost 1.1 billion data points created every week, with 2.5 billion GB of data being generated across the world. Likewise, about 500 GB of data is generated by offshore oil rigs and 100 GB of data from oil refineries per week. Also about 10,000 GB of data is generated by jet engines every 30 min. Overall, it is said that about 90% of the world's data has been generated in the last 2 years. Thus, when such a huge amount of data points and data are available on the public network like the Internet, they are susceptible to various attacks. Hence, it is essential to identify the possible security safeguards at the earliest.

With the growth of connected devices under IoT, there is an increase in the potential vulnerability on security, privacy, and governance. Though IoT can make people's life convenient, it might fail to ensure security and privacy of the user data leading to a number of undesirable consequences. For example, in 2015, IoT baby monitors were hacked through which the hackers were able to monitor the live feeds of the baby, change the camera settings, and authorize other users remotely to view and control the baby monitor [4]. During 2017, intruders could over-write the part of Ukraine's power grid that caused the first cyber attack [2]. Even the Internet-connected cars and wearable devices can also become a threat to the user's security and privacy.

In [5] Atmali et al. have analyzed the impact of the above attacks on IoT applications like power management, smart car, and the smart healthcare system. Through their study, they have projected that there is a need for security and privacy considerations at the level of (a) actuators, (b) sensors, (c) RFID tags, and (d) the Internet/ network. Attack on actuators in power management applications can lead to financial loss due to excessive power consumption. Similarly, in smart cars, these compromised actuators may control the broken system costing a driver's life. Also, in the healthcare system, these compromised actuators can inject the wrong dosage of medicine to a patient who is remotely monitored by the doctor. Likewise, a compromised sensor can fake the data that may lead to the wrong diagnosis of a patient. At the same time, these compromised nodes can reveal the personal information of the patient or the data related to a user's home through power management system.

Section 4.2 of this chapter explains the various security threats and attacks, followed by elements of security in Sect. 4.3, some of the lightweight existing solutions in Sect. 4.4, threat modeling tools in Sect. 4.5, Kali Linux-based ethical hacking in Sect. 4.5.4, major IoT security practices of E-IoT in Sect. 4.6, and lastly, conclusion in Sect. 4.7.

4.2 Security Threats and Attacks

Devices within the IoT communicate personalized data of many users. This data consists of user health information, bank details, passwords, location information, and many more. These devices are subjected to security threats like (a) malicious node in the network, (b) defective manufacturer, and (c) external adversary [5]. These threats lead to security attacks that can be initiated either by nature or human. The natural threats may include earthquake, floods, fire, and hurricane that cause severe damage to the computer system. Although it is hard to safeguard against natural calamities, it is advisable to reduce the damage by collecting backup of data through contingency plan. Accordingly, the security attacks caused by the humans affect the node privacy [6, 7]. Such attacks can be classified as follows [5, 8]:

1. *Information-level attacks:* All IoT devices are enabled with sensors that record the data from the physical environment and communicate the information over

the Internet. As the Internet is an open domain, attackers can easily tamper the information under following categories [7–9]:

(a) Denial of service (DoS): DoS is an attack over the network component that makes it unavailable for an intended user.
(b) Masquerade: An intruder behaves as an intended user and tries to talk with the network component.
(c) Modification of message: An intruder can alter or delete or fake a message sent by a legitimate user.
(d) Man-in-the-middle (MITM): MITM is a kind of attack wherein a malicious user takes control of the communication channel between two or more endpoints.
(e) Message replay attack: It is a security breach in which the message is stored by malicious node without the knowledge of intended users, and the malicious node transmits an altered message that is forwarded to the receiver.

2. *Adversary location attack:* An intruder can be present in any part of the IoT ecosystem. He can either be within or outside the IoT environment [9]:

(a) Internal attack: An attack caused by the components within the IoT border. It is also called as insider attack where the intruder tries to inject malicious code toward the IoT components.
(b) External attack: An attack caused by an advisory that is located outside the IoT environment in a remote place.

3. *Access-level attack:* Access-level attacks are broadly classified into active and passive attacks [10]. In the passive attack, an attacker can read the packet that is transmitted, but he/she cannot alter the packets like eavesdropping and traffic analysis. On contradictory, in active attack, the attacker sees the data and then alters the content of the data and transmits the altered data back to the network.
4. *Host-based attack:* Many devices in an IoT environment are made up of different manufacturers [10]. These devices are subjected to user compromise attack, software compromise attack, and hardware compromise attack. This is because the manufacturer can hold the devices' information which can be misused by him. Hence the production of such poorly secured goods results in compromising the user privacy. At the same time, any manufacturer can attack his competitors through their devices.

4.2.1 IoT Four-Layered Architecture and Associated Attacks

P. P. Ray [10] has surveyed various domain-based architectures that vary from RFID to healthcare to security to cloud services. But in general, a four-layered design of IoT is considered for different research as in Fig. 4.1. Mainly it comprises of perception layer, network layer, transport layer, and application layer. Each layer has its own properties and protocols. Primarily, the perception layer forms the physical

Fig. 4.1 Generic
four-layered IoT
architecture

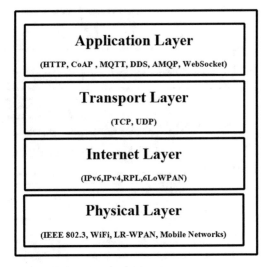

layer of the IoT ecosystem. It deals with sensors, devices, machines, actuators, and movements of unprocessed raw data. In this layer, the data transmission medium used is copper wire, coaxial cable, or radio wave. They have protocols like IEEE 802.3, Wi-Fi, LR-WPAN, 2G, 3G, 4G, and LTE networks [11].

Next is the Internet layer, which is also called the network layer. The main job of this layer is to provide host identification and packet routing. IETF has proposed many routing protocols that are suitable for low-powered device networks. Some of the protocols are IPv6, IPv4, RPL, 6LoWPAN, multipath RPL (MRPL) [12], energy-efficient probabilistic routing protocol (EEPR) [13], congestion avoidance multipath routing protocol (CA-RPL) [14], movement-aided energy balance (MABE) [15], least path interface beaconing protocol (LIBP) [16], and cognitive machine-to-machine RPL (CoRPL) [17].

Then comes the transport layer which is considered for end-to-end message transfer. The transmission can be either connection-oriented or connectionless with protocols like transmission control protocol (TCP) and user datagram protocol (UDP), respectively. This layer involves various processes like segmentation and reassembly of packets, congestion control, error control, and flow control.

Lastly, the application layer interfaces with all the lower layers by establishing a secure connection between the devices and servers. It uses standard port 80 and port 22 for most of HTTP and SSH protocols, respectively. Some of the protocols standardized by IETF are constrained application protocol (CoAP), message queuing telemetry transport protocol (MQTTP), extensible message and presence protocol (XMPP), data distribution services (DSS), and advanced message queuing protocol (AMQP) [11].

Likewise there are various attacks based on layers of IoT as shown in Fig. 4.2 [3, 18]. Sensing/perception layer is generally made up of sensors, RFIDs, NFCs, ZigBee, Bluetooth, and other intelligent hardware devices. These devices are exposed to more external attacks like node compromise attack, fake node injection,

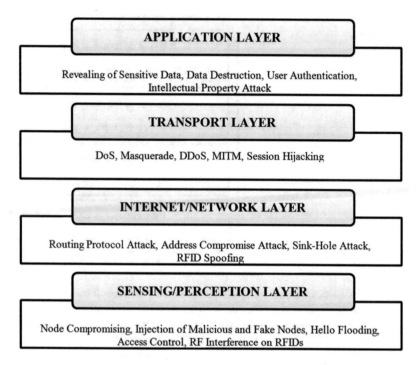

Fig. 4.2 Attacks based on architecture

access control, and RF interference on RFIDs. The second layer is the Internet layer and is subjected to attacks like address compromise attacks, routing information attack, RFID spoofing, and sinkhole attack. The next layer is the transport layer that experiences attacks like denial of service (DoS), masquerade, distributed DoS (DDoS), man-in-the-middle (MITM) attack, and session hijacking. And finally, the application layer experiences attacks like phishing attack, viruses, worms, malicious scripting, revealing of sensitive data, user authentication attacks, software vulnerability, and stealing of intellectual property.

4.2.2 Attacks Based on Phases of IoT

IoT can also be defined as an interconnection of "factual and virtual" objects placed across the globe that are attracting the attention of both "makers and hackers." IoT can be divided into five different phases as mentioned in [19] by Jeyenthi as shown in Fig. 4.3. The first phase is termed as the data collecting phase: primary interface between physical environment and sensors. There can be either static objects like body sensors or RFIDs or dynamic objects like sensors and chips on vehicles. The second phase is the storage phase: as many IoT devices are having low self-storage capability, IoT provides a server or cloud-based storage. Next is the intelligent pro-

PHASES	ATTACKS
Data Perception	Data Leakage, Data Authentication, Data Loss
Storage	Data Availability, Modification of Message, DOS, Attack on Integrity, Data Fabrication
Processing	Attack on Authentication
Transmission	Channel Security Attack, Session Hijacking, Routing Protocol Attack, Flooding
End-to-end Delivery	Man Induced or Machine Failure, Maker or Hacker

Fig. 4.3 Phases of IoT and their possible attacks

cessing phase: it is where the analysis of stored data and later appropriate services are provided to the users. IoT devices can be queried and controlled remotely using the results obtained after processing of data. The fourth phase is data transmission: it deals with processing of data communication between all of the above phases. Last is the delivery phase: it is where the activity of delivering the processed data to all the objects in time without being altered or hacked is performed.

Among the five phases, the data perception phase is subjected to more attacks like data leakage, data authentication, and data loss as the devices are easily available to users and hackers. Similarly, in storage phase, we can see attack on availability, modification of message, denial of service (DoS) attack, attack on integrity, and data fabrication. Attacks on authentication are seen at the processing phase, and channel security attack, session hijacking, routing protocol attack, and flooding are seen at the transmission phase. Lastly, at the delivery phase, man- and machine-made attacks are found as shown in Fig. 4.3.

4.3   Elements of Security

To ensure the IoT security, there are four elements of security [18]. They are device authentication, secure connections, secure data storage, and lastly, secure code execution. The device authentication grants the access privilege of the devices to the legitimate users. Secure connection enables the protection of the data that is travelling across the network (data in motion). Secure storage provides protection for data in rest using various lightweight encryption schemes. And lastly, the secure code execution serves the intended host machines to use the data and process it in a secure manner as in Fig. 4.4.

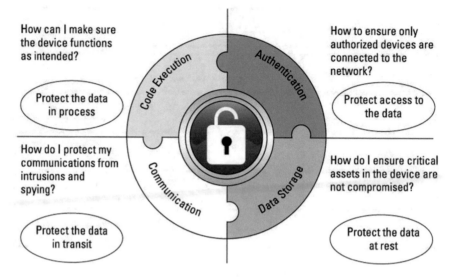

How can I make sure the device functions as intended?

Protect the data in process

How do I protect my communications from intrusions and spying?

Protect the data in transit

How to ensure only authorized devices are connected to the network?

Protect access to the data

How do I ensure critical assets in the device are not compromised?

Protect the data at rest

Fig. 4.4 Elements of IoT security [18]

As these poorly secured IoT devices can serve as means of entry point for cyber attackers by allowing various malicious individuals to re-program a device and cause malfunctioning, it becomes essential to provide security and privacy at the devices level. In order to develop a safer IoT solution, it is required to consider three major security requirements: (i) confidentiality, (ii) integrity, and (iii) authentication [20].

- Confidentiality means keeping information secret from the unauthorized user. For example, when transmitting certain sensitive data like location of military camp to the base station, it must be forwarded in secrecy to avoid intruders to understand the information that is being transmitted.
- Data integrity ensures that the messages transmitted are reached at the destination unaltered. Data integrity certifies the user that it has never been altered or corrupted by protecting the data over a communication channel.
- Authentication is a process of determining whether the data is transmitted by legitimate users or not. The user needs to identify the peer nodes that they need to communicate.

4.4 Lightweight Secure Measures for IOT

Elliptic curve cryptography (ECC) was introduced in early 1985 by Neal Koblitz and Victor Miller [21]. They stated that the hardness of ECC security depends on the discrete logarithmic problem defined on the elliptic curve. Later, Gura et al. [22] experimented ECC and RSA on an 8-bit CPU to compare their performance and found that the use of ECC for a lower-bit processor provides the same level of security as that of RSA. Later during 2013, Wenger [23] developed an ECC-based access control scheme over a prime field on 16-bit MSP430 micro-controller whereby the results confirmed the feasibility of ECC for resource-constrained devices.

Basically, ECCs are often implemented by using a static public elliptic curve that is shared among all the users in the network. In [24] the recommended elliptic curve domain parameters are provided for the Weierstrass curve equation $y2 = x3 + ax + b$ that is accepted by various researchers [25]. Liu et al. [26] have proposed software and hardware architecture for resource-constrained embedded devices. Their work has shown the feasibility of ECC on the embedded system. But the use of a fixed elliptic curve can be challenged on intensive cryptanalysis. Wang et al. [27] made a study on using a fixed prime field to build a crypto-system for applications developed for different processors varying from 8 bits to 256 bits.

A lightweight multi-message and multi-receiver heterogeneous-based signcryption is proposed by Rahaman et al. [28]. They have used the hybrid elliptic curve to generate signatures. The work is evaluated for various attacks like replay attack, forward secrecy attack, and unforgeability using the AVISPA simulator tool. For the heterogeneous environment, the attackers are inclined to impersonate legitimate users. To solve such an issue, Jingwei Liu et al. [29] have proposed a novel authentication scheme. They have provided a lightweight anonymous authentication and key agreement scheme as proposed. Their scheme could toggle between the public key infrastructure (PKI) and certificates analysis. Their method showed resistance against replay and DoS attacks.

The combination of cloud-based services with IoT has raised the issue of limitation regarding low latency and high mobility. To address such issues, Haldorai et al. [30] have proposed the authentication and key agreement scheme for fog-based IoT for the healthcare application. By using bilinear key agreement protocol, they have proposed a protocol that showed resistance against MITM, replay attack, known-session key attack, and intractability.

Recently, based on card shuffling logic, a data confidentiality algorithm is designed using ECC, proposed by Khan [31]. The use of random card shuffling has shown double encryption and increased the security of the algorithm. The algorithm can encrypt or decrypt any type of ASCII values. As the algorithm uses ECC, it is suitable for resource-constrained devices. Li et al. [32] proposed a lightweight mutual authentication protocol using public-key encryption schemes for smart city applications. Their simulated work has shown a balance among ciphertext size, usability, and efficiency. The generation of online and offline signatures created overhead on the device storage. Diro et al. [33] have used ECC to provide lightweight encryption for fog-based IoT applications. They have shown better efficiency regarding runtime, throughput, and ciphertext expansion. But they could only handle a smaller data size.

An OTP-based end-to-end authentication scheme was proposed by Shivraj et al. [34]. Their scheme used Lamport's OTP scheme with ECC-based authentication algorithm. Even though the scheme performed better than existing OTP-based signature schemes, they could not justify the implementation on a real-time scenario. A security framework for IoT and cloud computing is proposed by Daisy Premila et al. [20]. They used ECC-based message encryption and multi-factor authentication to ensure confidentiality, integrity, privacy, and authentication. They have concluded that the use of ECC-based security measures is better than RSA to eliminate the ambiguity and enhance security. But the research to collaborate IoT and cloud computing needs to depend on infrastructure.

During 2018, to address the usage of the static curve in ECC, Jia Wang et al. [35] proposed a dynamic elliptic curve-based Internet of Vehicles (IoV) network. Their work showed good computational efficiency and security for a smaller key size. But storing the elliptic curves as a plain text in embedded systems would lead to security concern. To address the data integrity issue of Java card-based application, Gayoso et al. [36] initiated the use of ECC-based encryption algorithm called an elliptic curve integrated encryption scheme (ECIES) and concluded that ECIES-based encryption is the best among encryption schemes for resource-constrained devices.

4.5   Threat Modeling for IOT Security

Threat modeling (TM), whose lifecycle is depicted in Fig. 4.6, is a process of identifying the potential threats, enumerating and prioritizing the threats, and providing countermeasures to mitigate the threats. TM can be applied to any platform of a working process like software, application, networks, IoT devices, or business processes. Shostack [37] has summarized the reasons to incorporate a threat model in SDL which are (i) to find the bugs at the earliest, (ii) understand the security requirements, and (iii) engineer and deliver a better product. Basically, TM includes components like *target-of-evaluation* (*ToE*) (a design or model of what type of platform needs to be analyzed), a list of *assumptions* that can be threats on ToE, a list of *potential threats* on ToE, *possible countermeasures* toward the identified threats, and *verification of success* (VoS) that validates the threat model.

Before modeling a threat, there are four questions that need to be answered, which are as follows:

1. What are we building? A detailed data flow diagram (DFD) is designed by specifying various roles and responsibilities of each participant.
2. What can go wrong? The various possible threats are analyzed using methods available in STRIDE, PASTA, STRIKE, or VAST.
3. What are we going to do about that? Potential mitigation strategies against the threats are framed.
4. Did we do a good job? Once the mitigation is applied, the system is validated for the stability and security against the threats.

4.5.1 Microsoft Security Development Lifecycle

Microsoft SDL was introduced during 2008 to ensure security and privacy considerations throughout all the phases of the development process. This helped developers to build highly secure software, address security compliance requirements, and reduce development cost. The core of Microsoft SDL is threat modeling. Threat modeling helps in shaping the application design and meeting the security objectives of the company by reducing the risk severity. The five major steps of threat modeling involve (Fig. 4.5) the following:

1. Defining security requirements: To understand the ecosystem of the device, i.e., analysis of the ToE by framing various use cases. In this process the external and internal assets are identified.
2. Creating an application diagram: Here a detailed data flow diagram of the proposed ToE is framed with appropriate trust boundaries and security requirements for each participant.
3. Identifying the threats: Microsoft TMT follows STRIDE-based threat modeling where the threats are identified. Potential adversaries are identified under four categories called remote software attacker, network attacker, malicious insider attacker, and advance hardware attacker.
4. Mitigating the threats: For the threat identified, relevant countermeasures are established.
5. Validating that threats have been mitigated: Finally, the verification of the threat model against the mitigation is performed to check the stability of the proposed system.

4.5.2 STRIDE Framework Methodology

It is important to develop a secure design for any software application or system. Failing to do so may cost about 30 times higher than estimated cost [38]. Hence threat modeling plays a vital role in the software development lifecycle. Among various threat modeling methods like STRIDE, PASTA, VAST, and STRIKE,

Fig. 4.5 Microsoft Security Development Lifecycle (SDL) using TMT

DEFINE

VALIDATE   DIAGRAM
THREAT
MODELING

MITIGATE   IDENTIFIED

Table 4.1 STRIDE threat model with associated security properties

Threats	Security property	Definition
Spoofing	Authentication	Unauthorized access Using another user's identity
Tampering	Integrity	Malicious modification Unauthorized information changes
Repudiation	Non-repudiation	Denying to perform action
Information disclosure	Confidentiality	Unprivileged user gains access and compromises the system
Denial of service	Availability	Denying services to valid users Threats to system availability and reliability
Elevation of privilege	Authorization	Exposure of information to individuals not supposed to access

STRIDE has taken a major share among the industrial development processes [39, 40]. STRIDE is developed by Microsoft as a part of their Security Development Lifecycle. STRIDE is an acronym for spoofing, tampering, repudiation, information disclosure, denial of service, and elevation of privilege [41]. The security properties and attack types associated with STRIDE are summarized in Table 4.1 [38].

4.5.3   Overview of Threat Modeling Tool (TMT)

Microsoft TMT is used to provide assistance in analyzing the design of a system or an application in order to check for security risks and provide solution for the threat found. Figure 4.4 displays the initial page of TMT when launched. This page has two partitions; the top part is used to create the threat model of the user's choice using the templates provided by the Microsoft, while the bottom part helps the user to customize his own template on the default Microsoft Security Development Lifecycle (SDL) template as in Fig. 4.6.

4.5.4   Kali Linux-Based Ethical Hacking

Kali Linux was developed by Mati Aharoni and Devon Kearns of Offensive Security and was mainly suitable for digital forensic and penetration testing under ethical hacking [42]. Kali Linux has approximately 300 hacking tools that are broadly categorized under information gathering, vulnerability analysis, wireless attacks, web application, exploitation tools, forensic tools, sniffing and spoofing tools, password attacks, maintaining access, reverse engineering, and hardware hacking tools. Among these, the most commonly used tools are Metasploit framework, dsniff, tcp-dump, Nmap, Wireshark, Aircrack-ng, Armitage, Burp Suite, BeEF, and so on [42].

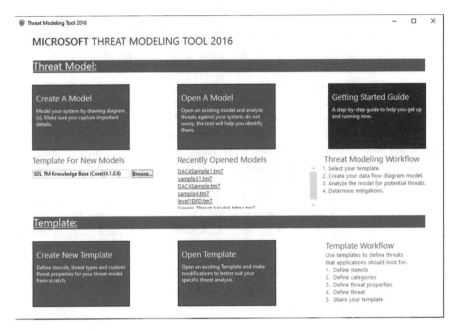

Fig. 4.6 Microsoft TMT initial screen

Featuring the rapid growth of smart cities, Barghuthi et al. [43] have made a study of how the increase in the population of smart cities shall add to an increase in the security breach and damage to businesses by 2050. Thus, they have proposed Kali Linux-based vulnerability assessment and penetration testing solution using low-cost Raspberry Pi 3 devices. Through their results, it has been concluded that Raspberry Pi 3 can be used as a machine to check the vulnerability similar to any traditional PC or laptop-based Kali Linux machine.

To replace the expensive and resource-intensive devices used for industrial vulnerability and assessment tests, Hu et al. [44] proposed an automated vulnerability assessment using OpenVAS and Raspberry Pi 3 device. They have detailed methods for analyzing the vulnerability assessment of distributed architecture. They made the study on variables like CPU temperature, CPU usage, and CPU memory of the device at the time of vulnerability assessment.

Visoottiviseth et al. [45] developed a GUI-based penetration testing tool called PENTOS used for IoT devices. PENTOS runs on Kali Linux and is specifically designed for the ethical hacking of wireless communication like Wi-Fi and Bluetooth. PENTOS enables the analysis of password attack, web attack, and wireless attack that ensure to gain access privilege of the various algorithms. They also have explained the Open Web Application Security Project (OWASP) specified ten vulnerabilities of IoT applications.

Finally, they have given the recommendations for the secure deployment of the IoT environment. Denis et al. [46] performed various penetration tests using tools available on Kali Linux. They were able to set up a private network and generate attack reports and visualize the reports using Kali Linux tools. The attacks they performed were hacking phones, MITM attack, smartphone penetration testing, spying, hacking phones' Bluetooth, and hacking WPA-protected access, and then they hacked the remote PC using IP and open ports.

Liang et al. [47] experimented on different methods of doing DoS attack using Raspberry Pi-based Kali Linux. They have provided an attack framework and compared various DoS attacks on their framework. They have used Hping3 with random IP, SYN flood with spoofed IP, and TCP connection flood tools. The comparison was made under the parameters like CPU utilization, memory utility, time for the success of an attack, and packet loss rate. Ryan Murray [48] has proposed a forward-looking approach for a secure eHealth solution called HealthShare. It could share data among various organizations that were hosting the patient's data over the cloud. Detailed steps as to conduction of MITM and DoS attack using tools like Ettercap, Pexpect, manual SET, threads using the timer and Nmap timer, and Scapy have also been provided.

4.6   Major E-IOT Security Practices

As E-IoT is deployed on a larger scale with heterogeneous business applications, the cybersecurity space has obtained an intense research spectrum. Some of the important security practices that should be followed by enterprise IoT are explained below.

(a) *Understand your endpoints*: Every endpoint of the business network is assembled by various manufacturers using different open-source operating systems. These devices are potential entry points for cybercriminals. Thereby, it is essential to deploy devices in a tamper-proof environment using secure hardware and software resources.

(b) *Track and manage the endpoints*: Business enterprise poses the responsibility of constant check on the devices that are deployed under their network and should be updated with frequent firmware and security patches. As it is infeasible to monitor each device physically, Earl Perkins of Gartner Solutions has recommended "rolling out an asset discovery, tracking, and management strategy" to be implemented before the IoT project begins.

(c) *Change the default passwords and other credentials*: The manufacturers set their devices with a common default password, which has to be updated by the enterprise officials frequently. This is because, most of the time, hackers are well aware of default passwords and sneak into your network by brute force attacks.

(d) *Execute risk-driven strategies*: IoT projects need to be analyzed for risk possibility using various threat modeling tools. Such tools help to identify the risks in the network and guide the network administrator to take corrective actions. Also, performing regular pen-testing at the hardware and software levels shall ensure the attack resistivity of the network.

(e) *Consideration of the latest encryption protocols*: Business enterprises should encrypt the data passing from and to their network using the updated and latest encryption schemes. If in case a single device is accessed by multiple users, then the focus should be on user authentication, identity-level control, and providing data integrity.

4.7 Summary

IoT is a rapidly growing network that has its major contribution in making the business enterprise smarter. E-IoT could connect to a diverse domain of applications and devices across the globe thus leading to various levels of attacks and threats. Various levels of hardware and software issues are studied with possible lightweight solutions. A generalized layer of security architecture is discussed, followed by a brief description on threat modeling tool. In addition, Kali Linux-based pen-testing on a real-time E-IoT is also studied. Finally, the major E-IoT practices are generalized that help future researchers to concentrate on the specific issues in E-IoT.

References

1. Telecommunication Standardization Sector of ITU. (2012). *Series Y. Global information infrastructure, internet protocol aspects and next-generation networks* (pp. 1–6). Geneva: ITU.
2. Vermesan, O., & Friess, P. (2014). *Internet of things-from research and innovation to market deployment* (Vol. 29). Aalborg: River Publishers.
3. Ahemd, M. M., Shah, M. A., & Wahid, A. (2017). IoT security: A layered approach for attacks and defenses. In *2017 international conference on Communication Technologies (ComTech)*, IEEE, pp. 104–110.
4. Khan, R., Maynard, P., McLaughlin, K., Laverty, D., & Sezer, S. (2016). Threat analysis of blackenergy malware for synchrophasor based real-time control and monitoring in smart grid. In *ICS-CSR*.
5. Atamli, A. W., & Martin, A. (2014). Threat-based security analysis for the internet of things. In *2014 international workshop on Secure Internet of Things (SIoT)*, IEEE, pp. 35–43.
6. Stankovic, J. A. (2014). Research directions for the internet-of-things. *IEEE Internet of Things Journal, 1*(1), 3–9.
7. AbdAllah, E. G., Hassanein, H. S., & Zulkernine, M. (2015). A survey of security attacks in information-centric networking. *IEEE Communication Surveys and Tutorials, 17*(3), 1441–1454.
8. Abomhara, M., & Kien, G. M. (2015). Cyber security and the internet-of-things: Vulnerabilities, threats, intruders and attacks. *Journal of Cyber Security, 4*, 65–88.
9. Nawir, M., Amir, A., Yaakob, N., & Lynn, O. B. (2016). Internet of things (IOT): Taxonomy of security attacks. In *3rd International Conference* on *Electronic Design (ICED)*, IEEE, pp. 321–326.
10. Ray, P. P. (2018). A survey on internet of things architectures. *Journal of King Saud University-Computer and Information Sciences, 30*(3), 291–319.

11. Arulmurugan, R., Sabarmathi, K. R., & Anandakumar, H. (2017). Classification of sentence level sentiment analysis using cloud machine learning techniques. *Cluster Computing, 22*(S1), 1199–1209.
12. Quynh, T. N., LeManh, N., & Nguyen, K. N. (2015). Multipath RPL protocols for greenhouse environment monitoring system based on internet of things. In *2015 12th international conference on Electrical Engineering/Electronics, Computer, Telecommunications and Information Technology (ECTI-CON)*, IEEE.
13. Park, S. -H., Cho, S., & Lee, J. -R. (2014). Energy-efficient probabilistic routing algorithm for internet of things. *Journal of Applied Mathematics*, Hindawi Publishing Corporation, Vol. 2014, Article ID 213106, 7 pages.
14. Haldorai, A., & Ramu, A. (2019). *Cognitive social mining applications in data analytics and forensics* (Advances in social networking and online communities). Hershey: IGI Global. https://doi.org/10.4018/978-1-5225-7522-1.
15. Haoru, S., Wang, Z., & An, S. (2013). MAEB: Routing protocol for IoT healthcare. *Advances in Internet of Things, 3*, 8–15. Scientific Research.
16. Ngqakaza, L., & Bagula, A. (2014, May 26–28). Least Path Interference Beaconing Protocol (LIBP): A frugal routing protocol for the internet-of-things. In *12th international conference proceedings, Wired/Wireless Internet Communications- WWIC2014*, Paris, France, pp. 148–161.
17. Aijaz, A., & Aghvami, A. H. (2015). Cognitive machine-to-machine communications for internet-of-things: A protocol stack perspective. *IEEE Internet of Things Journal, 2*(2), 103–112.
18. Miller, L., & Johnson, C. A. (Eds.). (2016). *IoT security for dummies*. Chichester: Wiley.
19. Anandakumar, H., & Umamaheswari, K. (2017). Supervised machine learning techniques in cognitive radio networks during cooperative spectrum handovers. *Cluster Computing, 20*(2), 1505–1515.
20. Bai, T. D. P., Rabara, S. A., & Jerald, A. V. (2015). Elliptic curve cryptography based security framework for Internet of Things and cloud computing. In *Conference on recent advances on computer engineering by WSEAS*, pp. 65–73.
21. Miller, V. S. (1985). Use of elliptic curves in cryptography. In *Conference on the theory and application of cryptographic techniques* (pp. 417–426). New York: Springer.
22. Gura, N., Patel, A., Wander, A., Eberle, H., & Shantz, S. C. (2004). Comparing elliptic curve cryptography and RSA on 8-bit CPUs. In *International workshop on cryptographic hardware and embedded systems* (pp. 119–132). Berlin/Heidelberg: Springer.
23. Wenger, E. (2013). Hardware architectures for MSP430-based wireless sensor nodes performing elliptic curve cryptography. In *International conference on applied cryptography and network security* (pp. 290–306). Berlin: Springer.
24. SEC, S. (2000). *SEC 2: Recommended elliptic curve domain parameters*. Standards for Efficient Cryptography Group, Certicom Corp.
25. Silverman, J. H. (2009). *The arithmetic of elliptic curves* (Vol. 106). New York: Springer.
26. Liu, A., & Ning, P. (2008). TinyECC: A configurable library for elliptic curve cryptography in wireless sensor networks. In *Proceedings of the 7th international conference on Information processing in sensor networks*, IEEE Computer Society, pp. 245–256.
27. Wang, J., & Cheng, L. M. (2017). Dynamic scalable ECC scheme and its application to encryption workflow design. In *Proceedings of the international conference on Security and Management (SAM)*, pp. 261–262.
28. Rahman, A. U., Ullah, I., Naeem, M., Anwar, R., Noor-ul-Amin, Khattak, H., & Ullah, S. (2018). A lightweight multi-message and multi-receiver heterogeneous hybrid signcryption scheme based on hyper elliptic curve. *International Journal of Advanced Computer Science and Applications, 9*(5), 160–167.
29. Liu, J., Ren, A., Zhang, L., Sun, R., Du, X., & Guizani, M. A. (2019). *Novel secure authentication scheme for heterogeneous internet of thing*. arXiv preprint arXiv:1902.03562.

30. Haldorai, A., Ramu, A., & Murugan, S. (2019). Smart sensor networking and green technologies in urban areas. In *Computing and communication systems in urban development* (pp. 205–224). Cham: Springer. https://doi.org/10.1007/978-3-030-26013-2_10.
31. Khan, M. A. (2018). Multidisciplinary Journal of European University of Bangladesh. *Cell, 1713*(006814), 01914–098494.
32. Li, N., Liu, D., & Nepal, S. (2017). Lightweight mutual authentication for IoT and its applications. *IEEE Transactions on Sustainable Computing, 2*(4), 359–370.
33. Diro, A. A., Chilamkurti, N., & Nam, Y. (2018). Analysis of lightweight encryption scheme for fog-to-things communication. *IEEE Access, 6*, 26,820–26,830.
34. Shivraj, V., Rajan, M., Singh, M., & Balamuralidhar, P. (2015). One time password authentication scheme based on elliptic curves for Internet-of-Things (IoT). In *5th National Symposium on Information Technology: Towards New Smart World (NSITNSW)*, IEEE, pp. 1–6.
35. Wang, J., Li, J., Wang, H., Zhang, L. Y., Cheng, L. M., & Lin, Q. (2018). Dynamic scalable elliptic curve cryptographic scheme and its application to in-vehicle security. *IEEE Internet of Things Journal, 6*(4), 5892–5901.
36. Gayoso Martínez, V., Hernández Álvarez, F., Hernández Encinas, L., & Sánchez, C. (2011). Analysis of ECIES and other cryptosystems based on elliptic curves. *Journal of Information Assurance and Security, 6*(4), 285–293.
37. Shostack, A. (2014). *Threat modeling: Designing for security*. Indianapolis: Wiley.
38. Verheyden, L. (2018). *Effectiveness of threat modelling tools*. Master thesis.
39. Bodeau, D., McCollum, C., & Fox, D. (2018). *Cyber threat modeling: Survey, assessment, and representative framework*. HSSEDI, The Mitre Corporation.
40. Meghanathan, N., Boumerdassi, S., Chaki, N., & Nagamalai, D. (2010, July 23–25). Recent trends in network security and applications. In *Third international conference, CNSA-2010, Chennai, India, 2010 proceedings*, Vol. 89. Springer.
41. Khan, R., McLaughlin, K., Laverty, D., & Sezer, S. (2017). Stride-based threat modeling for cyber-physical systems. In *2017 IEEE PES Innovative Smart Grid Technologies conference Europe (ISGT-Europe)*, IEEE, pp. 1–6.
42. K. Linux. (2016). *Kali linux tools listing*. https://www.kali.org
43. Al Barghuthi, N. B., Saleh, M., Alsuwaidi, S., & Alhammadi, S. (2017). Evaluation of portable penetration testing on smart cities applications using raspberry pi III. In *2017 fourth HCT Information Technology Trends (ITT)*, IEEE, pp. 67–72.
44. Hu, Y., Sulek, D., Carella, A., Cox, J., Frame, A., & Cipriano, K. (2016). Employing miniaturized computers for distributed vulnerability assessment. In *2016 11th International Conference for Internet Technology and Secured Transactions (ICITST)*, IEEE, pp. 57–61.
45. Visoottiviseth, V., Akarasiriwong, P., Chaiyasart, S., & Chotivatunyu, S. (2017). PENTOS: penetration testing tool for internet of thing devices. In *TENCON 2017–2017. IEEE region 10 conference*, IEEE, pp. 2279–2284.
46. Denis, M., Zena, C., & Hayajneh, T. (2016). Penetration testing: Concepts, attack methods, and defense strategies. In *2016 IEEE Long Island Systems, Applications and Technology Conference (LISAT)*, IEEE, pp. 1–6.
47. Liang, L., Zheng, K., Sheng, Q., & Huang, X. (2016). A denial of service attack method for an IOT system. In *2016 8th international conference on Information Technology in Medicine and Education (ITME)*, IEEE, pp. 360–364.
48. Murray, R. (2017). *A raspberry pi attacking guide*. Birmingham: Packt Publishing.

Chapter 5
Advanced Machine Learning for Enterprise IoT Modeling

N. Deepa and B. Prabadevi

5.1 Introduction

The Internet of Things (IoT) which is going to be the future enables every object (any living or non-living thing) to interact through the Internet without any form of human intervention like human-to-human or human-to-computer interactions. Ultimately like the Internet, IoT will also become a fundamental need. It is also termed as the Internet of Objects as it is an inter-connection of real-world entities [1]. Any object to be a part of the IoT system must have a unique identifier and capability to process the information. In terms of IoT, sensors and actuators allow every object in the IoT system to communicate. IoT makes everything smart; here, smart refers to making all the application-specific objects interact without any human intervention. Some of the IoT applications are smart cities, smart medicines, smart home, smart farming, and so on [2].

IoT makes activities smart by joining hands with various other technologies like embedded systems (supporting hardware and software communication), sensors (sensing live application-specific accurate information), big data analytics (managing colossal volume of data), cloud computing (data storage and scalability), and machine learning (making machines learn and predict from experiences). IoT has achieved its most significant impact on business through its variant named enterprise IoT (E-IoT). The term enterprise refers to a business that encompasses various front-office and back-office activities. IoT in convergence with business automates all the activities of an enterprise by embedding computing devices in enterprise objects thus enabling them to communicate for accomplishing the business

N. Deepa (✉) · B. Prabadevi (✉)
School of Information Technology and Engineering, VIT University,
Vellore, Tamil Nadu, India
e-mail: deepa.rajesh@vit.ac.in; prabadevi.b@vit.ac.in

© Springer Nature Switzerland AG 2020
A. Haldorai et al. (eds.), *Business Intelligence for Enterprise Internet of Things*,
EAI/Springer Innovations in Communication and Computing,
https://doi.org/10.1007/978-3-030-44407-5_5

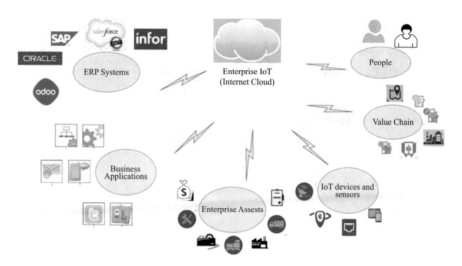

Fig. 5.1 Components of enterprise IoT

processes without or with less intervention of humans. This, in turn, will reduce manual operations thereby enhancing the efficiency of the business.

Enterprise resource planning (ERP) software is already in place for automizing all the business activities both front office and back end. Many ERP software such as Oracle, SAP, PeopleSoft, Fedena, Odoo, and CloudERP are providing better enterprise solutions to respective different business domains such as education, finance, manufacturing, and so on. So this will also be a part of an E-ToT environment. The emergence of E-IoT applications in the enterprise will automize the business processes. The various components of E-IoT are depicted in Fig. 5.1.

Though enterprise-specific business processes exist in each business, in general, most of the business processes respond based on the signals or commands from robust machines and other types of equipment. Also, these business processes can react based on the predefined business rules embodied in the E-IoT application. Fig. 5.2 depicts how the business processes interact with other entities and how the solution is retrieved by them. To deploy IoT applications on a large scale, the enterprises must have sufficient capital, good reach, enormous resources, and appropriate reasons for deploying it. Subsequently, it will be easier for businesses to attain greater business value, as well to utilize it for further adoption and business investment from the return of investment (ROI).

The greatest challenge for E-IoT is how the heterogenous business data will be handled. So, E-IoT bridges with technologies like mining to extract useful information from data stored and processed in the cloud and predicts different trends and patterns from knowledge gained through mining using machine learning and big data analytics. Also, real-time stream analytics have a more significant role to play with live data involved in business processes. Real-time streaming helps the enterprise decision makers to visualize data involved in every transaction and other business activities before taking an operational decision [3]. These technologies help in

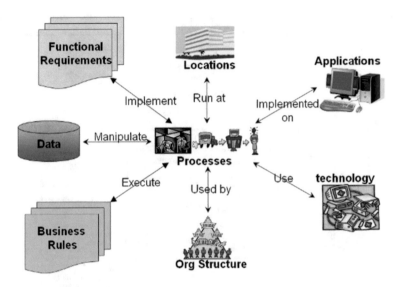

Fig. 5.2 Business process interaction with other entities in an enterprise

predictive maintenance of equipment, predicting market trends, determining production styles through customer buying pattern prediction, predicting the stock market, forecasting market demand to assist in demand-based manufacturing, predicting marketing trends, deciding prices for stock depending on different marketing stages of a product (i.e., from product promotion to stabilized brand product), and so on.

Integrating IoT devices with enterprise processes requires tiring work. This includes custom activity in various functionalities like hardware configuration, its deployment, and middleware configuration. So it is better to understand the existing business before proceeding for integration. Business process modeling (BPM) is the most popular technique in developing business models and accomplishing complex business processes in enterprises. Though several tools on BPM exist, those tools do not resolve the challenges in IoT-based business processes. Some of the challenges are adaptive event-driven business processes, distributed processes, and business processes that deal with unreliable data and resources [4].

5.2   Enterprise Forecasting

Business forecast refers to predicting the probable outcome in almost all the business processes or future situations of an enterprise and acting accordingly to attain success. Henceforth a better business forecasting facility of an enterprise determines its success in the short term. Most of the activities in the business are uncertain. So to be on the safer side and to react better in uncertain situations, forecasting

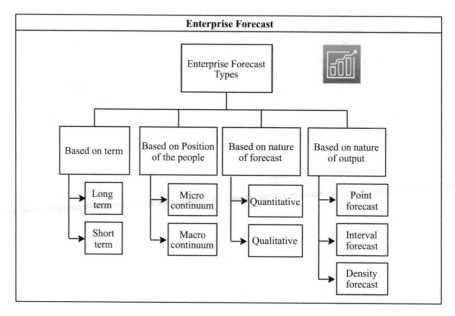

Fig. 5.3 Types of enterprise forecasts

is a must in all the activities we do. There are various models for enterprise forecasting [4] as depicted in Fig. 5.3. The models are classified based on term: long-term forecasts are those that determine the activities in the long run, whereas short-term forecasts determine the immediate future, i.e., short period. Therefore top management will be focusing on long-term forecast while intermediary level or basic level of management in an enterprise will be focusing on short-term forecast to decide their short-term plans.

The forecasts are classified based on the position or level of people in an enterprise hierarchy. At each level, management will focus on different stages of the micro- and macro-continuum. For instance, the production manager will be interested in forecasting the number of assembly parts, the number of employees, and the number of plants to be operated to complete the current tender while the top management will be bothered about the effective utilization of plants over a while (for a year). The forecasting can be classified in the way it is determined as qualitative or quantitative. The qualitative forecast is based on expert opinions or input from the enterprise's customers. Most enterprise practitioners will prefer this when having very meager or no information about past or historical data or when the market is facing a new scenario. In such situations, experts will help to predict the future based on prior experiences. Some of the examples of this approach are market research where predictions are done based on a survey with a certain group of people and Delphi approach where the experts are forced to give forecasts. The quantitative forecasting is just opposite to the former one discussed. It is merely based on accurate historical data and omits the human involvement in the analysis.

Thus predictions are done based on the previous data and not considering the expert opinions (i.e., manipulation based on human judgments). One of the commonly used quantitative approaches is the indicator approach which considers the relationship among certain indicators (the positive indicators) and determines the lagging one's performance by leading indicator's previous performance results. The other methods are econometric and time series. The econometric model is purely based on mathematical or statistical approaches like regression techniques such as simple linear regression, multiple regression, and regression with time-series data. Instead of relying on the static data members relationship, it will calculate the consistency of the variables in dataset over time and then determine the relationship strength. The time-series models generally refer to the set of models that predict based on past data. It differs from others by considering more weight for recent data and omitting the outliers. The enterprises over the other models best prefer this approach. The forecast can be classified based on the output as a point, an interval, or a density forecast. If the forecast output is the single best value, then it is a point forecast, whereas if a forecast output falls between a range of values, then it is an interval forecast, or if the output is probably distributed for future value, then it is a density forecast. Choosing one best forecast method among many available methods is determined based on various factors such as type of product (new vs. old), goals (short term or long term), and constraints (cost and time factors). However, it is infrequent that one single model fits for all the enterprise forecasts. Always a hybrid combination of forecasts will give the best results. In general, forecasting is done in five steps, namely, problem analysis and dataset collection, data preprocessing, building and evaluation of the model, implementation of the model, and evaluation of the results. Specifically, the performance of the forecast depends on the type of data and the source from which it was collected. Predicting the future can be better assisted with machine learning algorithms.

5.3 Machine Learning Algorithms and Their Application

Machine learning is a technology in which the computer will have the capability to learn from the available dataset and implement the learning to predict unique patterns from the data. It helps in developing models for data analytics. It identifies hidden knowledge or interesting patterns from the available dataset and predicts the future in most of the problems. The two popularly used methods in machine learning are supervised learning and unsupervised learning.

The supervised machine learning algorithms are allowed to perform learning on the dataset in which class labels or outcomes are known. These kinds of algorithms are used in credit card companies where the past fraudulent dataset is used for training the model and the learning helps to predict the fraud in the future. Some of the popularly used supervised algorithms are support vector machine (SVM), decision tree classifier, naive Bayes algorithm, linear regression method, logistic regression method, K-nearest neighbor algorithm, etc. The unsupervised machine learning

Fig. 5.4 The steps in the development of machine learning models

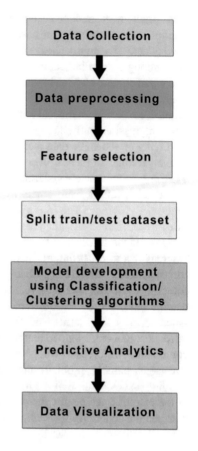

algorithms learn from the dataset in which class labels or outcomes are not known. The main aim of these kinds of unsupervised algorithms is to analyze the data and produce interesting hidden patterns in the data. For example, these algorithms can be used to group customers with similar purchasing behavior in retail marketing. Some of the unsupervised learning algorithms are hierarchical clustering algorithm, K-means clustering algorithm, self-organizing maps, etc.

The various steps in the development of machine learning models are shown in Fig. 5.4. The data relevant to the problem should be gathered either from real life or from past historical data. As the raw data collected may contain missing information or may be erroneous and inconsistent, statistical methods can be applied to clean the data. The raw data collected may be in a different unit of measurement, and it should be normalized to the same unit of measurement which helps to improve the accuracy of the developed model. If the dataset consists of high-dimensional data, feature selection/dimension reduction can be applied to reduce the dimension of the data or to identify the useful set of features. Then the dataset needs to be divided for training (learning) and testing purpose. The model is developed by applying appropriate classification or clustering algorithms. From the results, knowledge prediction is obtained to make reliable decisions for the identified problem. The results can be visualized with the help of graphs for further analysis and comparison with the state-of-the-art techniques.

As most of the enterprises are working with a large amount of data nowadays, machine learning algorithms have become popular in the business domain. They develop models using machine learning algorithms to identify the interesting patterns in their huge dataset. They make rapid decisions based on the knowledge acquired through machine learning models on their dataset which helps them to obtain maximum profit. Machine learning algorithms are applied to the dataset acquired from IoT sensor devices to predict the knowledge in various domains.

As there is a rapid increase in the application of IoT devices, the amount of data generated by them also increases. In order to process and analyze the data collected from IoT devices, models should be developed which can process the huge amount of data. Machine learning algorithms have proved to provide better results in processing data obtained from IoT sensors. Machine learning algorithms have globally become increasingly popular, and they have been providing various benefits in our day-to-day life. The various application domains of machine learning algorithms are shown in Fig. 5.5.

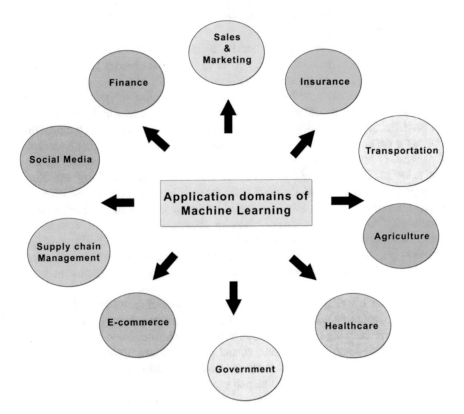

Fig. 5.5 Application domains of machine learning algorithms

Machine learning algorithms have been applied to develop classification and prediction models for the development of agriculture such as land suitability analysis and crop classification based on multiple parameters [5–11]. A new disease risk prediction algorithm based on convolutional neural network was developed to predict the chronic diseases in patients. Experimentation was done using the hospital data collected from China. Missing data in the dataset was handled by latent factor model. The results of developed algorithm were compared with existing methods and showed 94.8% prediction accuracy [12]. A framework was developed to detect the fake and abusive records in the claims received from the customers in healthcare insurance agencies. In the framework, pairwise comparison using analytical hierarchy process method was applied for the calculation of weights of criteria involved. Expectation maximization was used for clustering similar records. The developed framework was validated using a real dataset with six different fraudulent records [13].

Supervised learning algorithms were used to identify the optimal land location for starting a new retail store. The real dataset collected from New York was used for studying the prediction accuracies of various machine learning methods [14]. A non-linear non-parametric forecast model was built using machine learning algorithms such as linear regression to predict the credit risk of the consumers of a commercial bank [15]. A model was developed to classify and predict the claim amount from the automobile insurance agency.

Machine learning algorithms such as the hierarchical clustering method, heuristic method, and regression algorithm were used to build the model. Clustering algorithms were applied to group the policyholders with similar factors, and the regression method was used for classification and prediction of claim amount [16]. A machine learning model was developed using an enhanced AdaBoost algorithm for classification and prediction of sales data in the retail business. The model processes the data obtained from daily transactions and utilizes it to predict future sales for inventory management. Accordingly, proper decisions can be made to improve the sales of the retail business [17]. A study was conducted by the Korea Transportation Safety Authority to find the most suitable parameters that affect the safety measures of the agencies, and based on that, a strategy for promotion can be suggested. Cluster-based negative binomial regression method was used for the selection of factors such as socioeconomic factors, demographic factors, conditions of roadways, driver behavior, and traffic rule violations. Negative binomial regression models were built to identify the unique parameters for the prediction of deadly crashes [18].

A machine learning model was built for the prediction of rainfall under the hydrological research with non-linear pattern of data which is very difficult to process. Machine learning algorithms such as recurrent artificial neural network algorithm, support vector machine method, and particle swarm optimization approach were used to develop the prediction model. Recurrent artificial neural network algorithm was used to predict the rainfall, and support vector machine is used to find the solution for the non-linear regression and the problems in time-series data. Particle

Table 5.1 Applications of machine learning algorithms in different domains

S. no.	Purpose	Technologies/methods used
1	To assist farmers to take decision on agriculture crop cultivated in their land for agriculture development	Machine learning algorithms [5–8]
2	Multi-class classification model for agriculture land suitability analysis	Multi-layer perceptron, IoT sensors [11]
3	Prediction algorithm to forecast the chronic diseases in patients	A convolutional neural network, latent factor model [12]
4	A framework for the detection of fake and abusive insurance records for the claims submitted by the customers	Analytical hierarchy process [13]
5	A model to find the optimal site location for starting a new retail shop	Machine learning algorithms [14]
6	Non-linear non-parameteric forecast model to predict the credit risk of bank customers	Linear regression method [15]
7	A decision model to classify and predict the insurance claim amount for automobile insurance agency	Hierarchical clustering method, heuristic method, and regression algorithms [16]
8	A model to classify and predict the sales data for a retail business	AdaBoost algorithm [17]
9	A model to find suitable attributes for safety measures in transportation agencies	Cluster-based negative binomial regression method [18]
10	A machine learning model to predict the rainfall with non-linear pattern of data	Recurrent artificial neural network algorithm, support vector machine, and particle swarm optimization approach [19]

swarm optimization method is used to select the relevant parameters for the development of the model [18]. The survey of applications of various machine learning methods in different problems is summarized in Table 5.1.

5.4 Applications of Machine Learning in E-IoT

From the past few years, the costs of IoT sensors and chips used to store the data have been diminishing. The data collected by IoT sensors are very huge in the cloud and nowadays it is very fast also. In order to analyze the data stored in IoT, machine learning plays a major role. This can be possible by integrating the physical devices, cloud, and machine learning models which in turn will increase the performance and efficiency of decisions taken in the business enterprise [20]. Enterprise IoT is becoming popular due to the technological advancement where the physical sensor devices are integrated with software modules to take rapid and accurate decisions in business applications. Enterprise IoT reduces manpower and helps to improve the overall productivity of businesses. The physical devices are interconnected with the

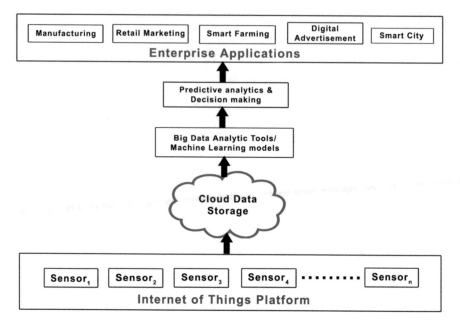

Fig. 5.6 Applications of machine learning models in E-IoT applications

Internet thus transforming the required data to the cloud for data analysis which helps to improve the enterprise.The architecture diagram depicting the application of machine learning models in enterprise IoT is shown in Fig. 5.6.

IoT devices have been recently used in the manufacturing divisions of industries where the data collected from various sensors are used for predicting some useful information. A predictive model was developed using autoregressive integrated moving average method. The data collected from various sensor devices fixed in a slitting machine were applied to the developed predictive model in order to forecast the time-series data. Thus machine learning has proved its importance in enterprise IoT [21]. A machine learning model was built using multistage meta-classifier for network traffic analysis. This model classifies and identifies the IoT devices which are connected to a specific network. The model initially distinguishes between the traffic data generated by the IoT devices and non-IoT devices. Later it classifies and identifies the correct IoT device from the chosen nine IoT devices. The model has proved to provide an accuracy of 99.28% [22].

A multi-class model was built using a supervised machine learning algorithm, namely, random forest, for the identification of fake IoT devices connected to the organization network. The model extracts the traffic data from the network to identify the type of IoT device from the given list of devices. In order to train the classification model, the traffic data collected from 17 IoT devices have been used with 9 class labels. The multi-class model has given 99.4% accuracy in detecting fake IoT devices [23]. The water crisis is a major problem nowadays for sustainable agriculture development. There is a need for the effective utilization and management of water. A machine learning-based model was developed using multiple linear regression algorithm for monitoring the water level in the agriculture field. IoT devices were fixed in the irrigation field which will continuously collect details about the underground water level and other details such as sunlight, air pressure, rainfall, temperature, etc.; this will help the farmers to get details about the water scarcity which will hit the land in the future [24].

A model was developed using hidden Markov method to predict the disease in grapefruits and give an alert to the farmers to apply pesticides through SMS to their mobile phones. In order to develop the model, various sensors to collect details such as moisture, humidity, temperature, and leaf wetness were fixed for data collection [25]. A model was built using a machine learning algorithm to alert about the health condition of a person. The heartbeat of the person was measured using IoT sensor and taken as a parameter to predict the stress behavior of the person in this model [26]. A business intelligent system was proposed to track the emotion and behavior of the customers using IoT devices which runs on Apache Spark cluster in the retail industry. An intelligent trolley, namely, EmoMetric, was developed to provide insights on customer behavior by using the customer's face. The results were compared with other popular techniques and showed 95% accuracy [27].

A classification model was developed using a supervised neural network algorithm, namely, multi-layer perceptron, to classify the threats in IoT sensor network. The results of the developed model produced 99% accuracy and proved to give better solution in the detection of distributed denial-of-service attacks [28]. A framework was presented for the development of the decision support system to work inside the ecosystem of IoT. Network communication data related to quality for electric smart meter is analyzed to make decisions on whether a technician needs to be sent to the customer address to solve the issue in the electric smart meter. The framework was implemented using a Bayesian network, and the results were compared with the results obtained from naive Bayes, decision tree, and random forest algorithms [29].

An artificial intelligence-based bot, namely, SamBot, was developed for a marketing purpose which is to provide interactive answers to customers' questions in the corporate website of Samsung IoT. SamBot is integrated with the marketing domain knowledge such as frequently asked questions, product promotions, etc., Sometimes the bot gives random answers if the knowledge is not available. In order to increase the efficiency of the bot, a supervised machine learning algorithm was applied to improve the knowledge of the developed SamBot [30]. A big data-based cloud system architecture was proposed to provide recommendations on a list of products to the customers. The architecture includes cloud to store the data relevant to products and purchases ranking. It analyzes the big data from the private and hybrid cloud for ranking the products to improve the retention in marketing [31].

An integrated system model was developed to monitor the environment and climatic changes. The system integrates cloud platform, Internet of Things, remote sensing, global positioning system, geographic information system, and artificial intelligence. Web services, various sensors, and private and public networks have been used in this model to collect the data from various sources and transport it to the destination network. IoT-based data collection is done using multiple sensor devices, satellites, radar, balloons, meteorological instruments, mobile devices, Bluetooth, ZigBee, Wi-Fi, and RFID. Decision making and planning is scheduled in the application layer of the developed system architecture. The main functionalities of this layer are the storage, organization, processing, and distributing of the data related to climate and environment which are acquired from various sensor devices. And it also monitors ecological factors, pollution, resource management, and weather and forecasts disaster. It accomplishes the task of decision making related to the environment and weather conditions. A case study was conducted by collecting data from China to prove the efficiency of the system developed [32].

Internet of Things-based production system for processing electrical and electronic equipment waste was developed for a particular type of manufacturing unit for recycling purpose. In this system, a cloud-based architecture was implemented to provide services for the production department. The cloud platform is deployed to deliver computing units, software products, and storage space to the destination through the network. In the manufacturing division of an industry, it is not possible to provide resources such as tools, materials, and machines through the network. Hence Internet of Things is essential to connect and integrate the physical devices to the cloud platform and named as the Internet of Manufacturing Things (IoMT). This IoMT is used to exchange the resources and data but in various formats. Therefore there is a necessity to build an integrated method to assist the industrial information system among various service shareholders. In the cloud platform, electrical and electronic equipment waste can be combined with industrial informatics through IoMT [33].

A study is conducted on how to provide service through the development of the Internet to the manufacturing division of an organization by deploying Internet of Things platform. The study also analyzes how various benefits of the Internet of Things can be applied to improve product-service systems. Three case studies have been taken in which the Internet of Things was successfully implemented in three manufacturing companies in different sectors of industry such as power distribution and generation and metal processing units. The empirical results provided an idea to create different values of providing services to the industrial sectors in the manufacturing division [34].

The various enterprise applications of machine learning and the Internet of Things are summarized in Table 5.2.

5.5 Issues in Enterprise Internet of Things

There are several issues in integrating Internet of Things with enterprise. The major challenges faced by enterprise IoT are shown in Fig. 5.7 [35].

Table 5.2 Application of machine learning algorithms on enterprise Internet of Things

S. no.	Purpose	Technologies/methods used
1	The predictive model to forecast time-series data	IoT, autoregressive integrated moving average method [22]
2	Machine learning model for network traffic analysis and to distinguish between IoT and non-IoT devices	Multistage meta-classifier, IoT [23]
3	Multi-class model to identify fake IoT devices in organization network	Random forest algorithm, IoT sensors
4	A decision model to monitor the groundwater level in sustainable agriculture development	Multiple linear regression algorithm, IoT sensor devices [24]
5	A model to predict the disease in grapefruits and recommend suitable pesticides to the farmers	IoT sensor devices, machine learning algorithm, namely, hidden Markov method [25]
6	An approach to predict the stress behavior of a person	Machine learning algorithms, sensor devices [26]
7	Business intelligent system to predict the emotion and behavior of the customers	IoT devices, Apache Spark cluster, machine learning methods [27]
8	A classification model to classify the threats in IoT sensor networks in distributed denial-of-service attacks	Multi-layer perceptron, IoT sensors [28]
9	A decision support system to provide technical support for smart electric meter	Internet of Things, Bayesian classifier, decision tree, random forest [29]
10	Artificial intelligence-based SamBot to provide online help for Samsung customers	The supervised machine learning algorithm, IoT devices [30]
11	A recommendation model using cloud system architecture to give product suggestions for the customers	Cloud platform, IoT, machine learning algorithms [31]
12	An integrated model to monitor the environment and climate change	IoT, multi-sensor devices, cloud platform, artificial intelligence, remote sensing, global positioning system, geographic information system, web services [32]
13	IoT-based production system to process electrical and electronic equipment waste in manufacturing division for recycling purpose	Cloud platform, IoT, machine learning algorithms, Internet of Manufacturing Things [33]
14	A system to provide services to the manufacturing division of the organization with the implementation of the Internet of Things	Internet of Things, cloud platform, machine learning algorithms [34]

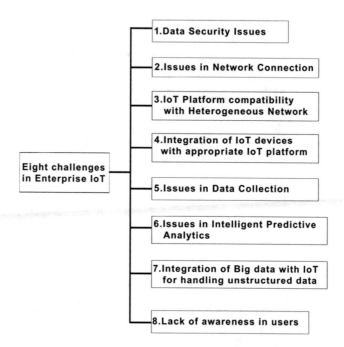

Fig. 5.7 Major challenges in integrating Internet of Things with enterprise

5.5.1 Data Security Issues

It has been found that most of the Internet attacks are targeted toward IoT devices. Topmost enterprises also face these security issues. Both users and enterprise have to face the problem of issues in privacy and security. Hence proper measure needs to be taken to enhance the privacy and security solutions in enterprise Internet of things.

5.5.2 Issues in the Network Connection

The major challenge in Internet connectivity is it is not possible to utilize the Internet facility at the same speed in all the places. It has been revealed that one of the major challenges in IoT implementation is the Internet connectivity issue. When IoT devices are fixed in remote locations, it is not possible to transmit the data with high speed. There may be disruption in Internet connection. Furthermore, based on the router, the signal quality collected by various sensors which transmit data to the network differs. The networks should be connected with the latest technologies to enhance the speed and quality of data transmission in the IoT platform. Moreover, the number of IoT devices connected to the network should also be monitored to obtain good network coverage.

5.5.3 IoT Platform Compatibility with Heterogeneous Network

IoT sensors and network are the major components in implementing IoT. Not every computer connected to a network is capable of connecting to advanced sensors. And also not every network has the capability to communicate and share the data effectively. Different sensors have the distinct power-consuming capability and may not be able to produce the same results. It is mandatory to deploy the latest technologies to integrate the different sensors to the heterogeneous network which will communicate data in the same manner.

5.5.4 Integration of IoT Devices with Appropriate IoT Platform

In order to incorporate the IoT devices with enterprises, the IoT physical devices should be integrated to correct IoT platforms. If it is not properly connected to appropriate IoT platform, functional abnormalities and inefficiency in the delivery of solutions to the customers may happen. It has been found that nearly half of the implementation cost should be spent for the integration of IoT devices with appropriate back-end systems and IoT platforms. Only then the enterprises can explore the data through big data tools and predict the knowledge. Therefore appropriate IoT devices and platform for integration of application-specific data should be devised before integration. This will help in avoiding the issues.

5.5.5 Issues in Data Collection

The main purpose of integrating IoT devices with the enterprise is to capture the data from various physical devices, convert to some standard format, apply machine learning models, and predict meaningful information for business purposes. If there are anomalies in the real environment where IoT devices are fixed, say for example power shutdown or problems due to natural calamities, wrong data may be collected by the IoT sensor devices.

5.5.6   Issues in Intelligent Predictive Analytics

The major contribution of enterprise Internet of Things is that it transforms the data to supply knowledge. If there are flaws in the collected data, the prediction capabilities will not be correct. In most of the places, IoT stakeholders find it difficult to collect correct data, and they confirm that the analytics made by them are not flexible and strong. Therefore the latest analytic software like machine learning models can be developed to check the integrity and consistency of the data before intelligent predictive analytics can be made.

5.5.7   Integration of Big Data with IoT for Handling Unstructured Data

As there is an increase in connected devices, the problems in handling unstructured data also increase. The other major challenge for the enterprises is to decide which data is useful because only valuable data can be used for data analytics. The data collected from various physical devices consists of unstructured data. Unstructured data cannot be stored in SQL databases. When unstructured data is saved in NoSQL format, it is a little complex to access the data. Big data frameworks such as Cassandra and Hadoop reduce the complexity of handling unstructured data. But the integration of IoT with big data is itself a major challenge.

5.5.8   Lack of Awareness in Users

Many customers use mobile apps, but they are aware of the Internet of Things incorporated in it. Even though it is not necessary for the consumers to have technical awareness about IoT devices, the lack of IoT knowledge may lead to hesitation with respect to cost and security of such devices. And that may slow down the usage of such technologies. In a survey conducted by Cisco, it has been found that 53% of customers do not want their data to be collected from their devices. This shows the lack of awareness about the IoT platform among the consumers.

5.6   Advanced Machine Learning Techniques in E-IoT

As discussed priorly, machine learning assists in various business activities and allows the decision makers to make a wise decision in different aspects of business solutions. Based on the historical data, the machines are trained and then test data is given to test the system's behavior without human intervention. In this section, we provide a brief overview of advanced machine learning techniques for E-IoT.

5.6.1 Application of Machine Learning for Demand Forecast in Enterprise Supply Chain

One of the major activities in an enterprise is a business forecast. The supply chain is the chain that connects the supplier from whom raw material is procured, the manufacturer who produces the product from raw material procured, the dealer, the wholesaler, the retailer, and finally the end buyer or the user. To stay connected and to retain good relationship with various roles in this chain, an enterprise must make wise decisions in all the activities in this chain. A supplier needs to forecast manufacturer demands, and in turn, a manufacturer must forecast various other processes in predicting the customer. Although customer demand is predictable, the manufacturer's demand will fluctuate randomly. So the biggest challenge is predicting manufacturer's demand in this collaborative forecasting environment. It has now become mandatory for an enterprise to realize the integrated information across various stakeholders in its supply chain. This determines the value of an enterprise. Fig. 5.8

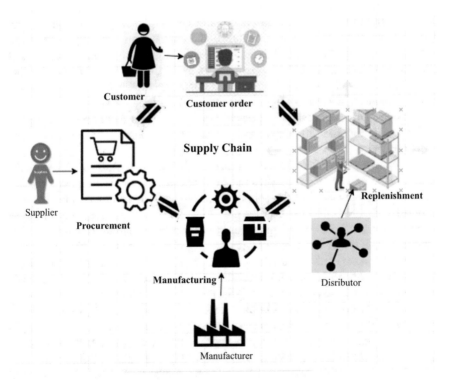

Fig. 5.8 A supply chain

Fig. 5.9 Bullwhip effect
and distorted
demand signal

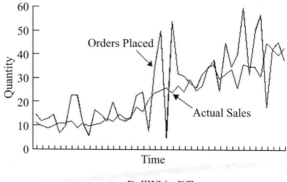

a.   BullWhip Effect

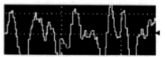

b.   Distorted Demand Signal

shows various stakeholders in a supply chain. A major challenge in forecasting the demand of a manufacturer is the demand signal distortion through the supply chain. This distortion may be due to inaccurate information passed into the supply chain [36]. The accurate result on forecasting can be attained through supply chain collaboration, i.e., by involving all the stakeholders of the supply chain. Since it is not always possible to involve all the participants every time, distorted demand signal should be observed in case of full collaboration and partial collaboration. The distorted demand signal and bullwhip effect in manufacturing are shown in Fig. 5.9.

Though various forecasting techniques can be used as mentioned in the prior section, it is important to analyze these techniques to observe which induces the bullwhip effect.

The bullwhip effect is a distribution channel phenomenon. Bullwhip effect on the supply chain will be observed when there are changes in customer demand causing the companies involved in the supply chain to make more orders until the new evolved demands of the customer are met [37]. In a multi-level supply chain, if we move from right to left, that determines the flow of information for customers; on other side, it is the demand flow as shown in Fig. 5.8 (Fig. 5.9).

R. Carbonneau et al. had carried out an analysis of various advanced ML techniques for supply chain demand forecast and compared them with traditional basic methods that can be used for forecast [38]. The advanced ML techniques considered are neural network (NN), feed-forward error backpropagation NN, support vector machine (SVM), and recurrent neural network (RNN). The basic forecast techniques used are naïve forest, moving average method, average method, trend analysis, and multiple linear regression. They observed that machine learning algorithms provide better accurate forecasts than traditional methods. They suggest that trend analysis and naïve forest for demand signal processing show more errors in forecasts. Also Bontempi et al. suggest that the machine learning algorithms like artificial neural networks (ANNs), decision trees, nearest neighbor, and SVM outperform the traditional models such as Box-Jenkins and linear regression for time-series forecasting [39].

5.6.2 Application of Machine Learning for Enhancing Customer Value in E-IoT

It is a well-known fact that "a product with good features and excellent performance but with no customer is equivalent to a failed product." So a successful enterprise is one with valued customers. Therefore, customer value is more important in any enterprises. Some of the IoT applications focus on enhancing customer value in an enterprise. Understanding the mandatory need for customer value in an enterprise is considered to be a requirement for IoT adoption in an enterprise. Based on research, three categories of IoT applications were identified to enhance the customer value. They are monitoring and controlling, information sharing and collaboration, and big data analytics and business analytics [40]. Monitoring and control is dealt with handling data maintenance. It handles data collection of equipment's operational condition and its environment. Decision makers can track and control the performance of the equipment at any time from any place. I. Lee and K. Lee clarify how these three categories assist in attaining customer value [41]. Based on how the system is performing and how the customer is reacting to various operations of the pieces of equipment in the IoT environment, enterprises can decide how they can enhance their customers' needs. As IoT is connected with more number of devices, an enormous amount of data need to be collected from various actuators and sensors. These data will be transmitted to data analytic and business intelligence tools to make them appear simple for decision makers. These data can be used for forecasting customer demands in the market and their behavior. The last category of information sharing and collaboration helps in creating a better forecast. As discussed regarding collaboration in the supply chain, here information sharing happens through people to device, device to device, and people to people. So, collaboration among all the entities participating in the communication must be considered. When all the people/entities are connected collaboratively via IoT devices, here communication will be transparent and tasks will be accomplished on time. Bose and Mahapatra have done a brief survey on various applications utilizing machine learning algorithms in business data mining [42]. Ampazis had tried to predict the demand of customers in the basic and multi-level supply chain using machine learning algorithms [43]. He used trained ANN and SVM for regression. He found less uncertainty while using these advanced machine learning algorithms on three different huge datasets, namely, Netflix (to predict demand of movie rentals during the holiday period), Rotten Tomatoes (a movie review aggregator), and Flixster (movies-based social network). He has also suggested a computational intelligence approach for supply chain demand prediction [44].

5.6.3   Application of Machine Learning for Various Other Activities in E-IoT

There is no enterprise without a customer. Similarly, business partners are also equally important as customers. So Mori et al. have used machine learning algorithms for predicting best business partners using their enterprise profiles and their transactional dealings with various enterprises [45]. To predict the customer-supplier relationship, they have to build a machine learning-based prediction model. They suggest that SVM performs well in modeling the customer-supplier relationship. So this approach can be employed in E-IoT architecture to find plausible business partners based on their employee count, ranking, and foundation date and to better classify the customer-supplier relationships. Cumby et al. have proposed a machine learning-based prediction model for determining customer's shopping list from PoS purchase data. They have used decision trees (C4.5) for predicting class labels, and for learning each class, they have applied perceptron, winnow, and naïve Bayes [46]. They conclude that their tool can be used to build better customer satisfaction and can increase the revenue to up to 11%. So this can be employed in E-IoT for enhancing customer satisfaction.

To retain the customer, better customer relationship management is mandatory in any enterprise. Buckinx et al. proposed a multi-linear regression model to predict customer loyalty based on the transactional data available [47]. They have compared their model with other machine learning algorithms like random forest and relevance determination neural network and obtained better prediction rates from their models. So this model can be employed for E-IoT for predicting their customer loyalty.

5.7   Summary

This chapter discusses about enterprise IoT and its uses, challenges and issues, components, and applications in the real world. Various technologies that can be used for assisting E-IoT to sort various issues and challenges of E-IoT were presented. Brief details on machine learning and its importance, functionality, applications, and various algorithms had been carried out. Specifically how E-IoT is benefitted with the help of machine learning algorithms was addressed. As a concluding note, machine learning algorithms serve better with best accuracy results than traditional models existing for prediction of demand in supply chain, customer value, financial transaction, and business partners.

References

1. Xia, F., Yang, L. T., Wang, L., & Vinel, A. (2012). Internet of things. *International Journal of Communication Systems, 25*(9), 1101.
2. Gharami, S., Prabadevi, B., & Bhimnath, A. (2019). Semantic analysis-internet of things, study of past, present and future of IoT. *Electronic Government, an International Journal, 15*(2), 144–165.
3. Haller, S., & Magerkurth, C. (2011). The real-time enterprise: Iot-enabled business processes. In *IETF IAB workshop on interconnecting smart objects with the internet*, pp. 1–3.
4. Hanke, J. E., Reitsch, A. G., & Wichern, D. W. (2001). *Business forecasting* (Vol. 9). Prentice Hall: Upper Saddle River.
5. Haldorai, A., & Kandaswamy, U. (2019). Cooperative spectrum handovers in cognitive radio networks. In *EAI/Springer innovations in communication and computing* (pp. 1–18). Cham: Springer. https://doi.org/10.1007/978-3-030-15416-5_1.
6. Deepa, N., & Ganesan, K. (2019). Decision-making tool for crop selection for agriculture development. *Neural Computing and Applications, 31*(4), 1215–1225.
7. Deepa, N., Ganesan, K., & Sethuramasamyraja, B. (2019). Predictive mathematical model for solving multi-criteria decision-making problems. *Neural Computing and Applications, 31*(10), 6733–6746.
8. Haldorai, A., Ramu, A., & Murugan, S. (2019). Social relationship ranking on the smart internet. In *Computing and communication systems in urban development* (pp. 141–159). Cham: Springer. https://doi.org/10.1007/978-3-030-26013-2_7.
9. Deepa, N., Ganesan, K., Srinivasan, K., & Chang, C. Y. (2019). Realizing sustainable development via modified integrated weighting MCDM model for ranking agrarian dataset. *Sustainability, 11*(21), 6060.
10. Deepa, N., Srinivasan, K., Chang, C. Y., & Bashir, A. K. (2019). An efficient ensemble vtopes multi-criteria decision-making model for sustainable sugarcane farms. *Sustainability, 11*(16), 4288.
11. Vincent, D. R., Deepa, N., Elavarasan, D., Srinivasan, K., Chauhdary, S. H., & Iwendi, C. (2019). Sensors driven AI-based agriculture recommendation model for assessing land suitability. *Sensors, 19*(17), 3667.
12. Chen, M., Hao, Y., Hwang, K., Wang, L., & Wang, L. (2017). Disease prediction by machine learning over big data from healthcare communities. *IEEE Access, 5*, 8869–8879.
13. Kose, I., Gokturk, M., & Kilic, K. (2015). An interactive machine-learning-based electronic fraud and abuse detection system in healthcare insurance. *Applied Soft Computing, 36*, 283–299.
14. Karamshuk, D., Noulas, A., Scellato, S., Nicosia, V., & Mascolo, C. (2013, August). Geospotting: Mining online location-based services for optimal retail store placement. In *Proceedings of the 19th ACM SIGKDD international conference on knowledge discovery and data mining*, ACM, pp. 793–801.
15. Khandani, A. E., Kim, A. J., & Lo, A. W. (2010). Consumer credit-risk models via machine-learning algorithms. *Journal of Banking & Finance, 34*(11), 2767–2787.
16. Yeo, A. C., Smith, K. A., Willis, R. J., & Brooks, M. (2001). Clustering technique for risk classification and prediction of claim costs in the automobile insurance industry. *Intelligent Systems in Accounting, Finance & Management, 10*(1), 39–50.
17. Von Kirby, P., Gerardo, B. D., & Medina, R. P. (2017). Implementing enhanced AdaBoost algorithm for sales classification and prediction. *International Journal of Trade, Economics and Finance, 8*(6), 270–273.
18. Jung, S., Qin, X., & Oh, C. (2019). Developing targeted safety strategies based on traffic safety culture indexes identified in stratified fatality prediction models. *KSCE Journal of Civil Engineering, 2019*, 1–8.
19. Hong, W. C. (2008). Rainfall forecasting by technological machine learning models. *Applied Mathematics and Computation, 200*(1), 41–57.

20. Patel, P., Kaulgud, V., Chandra, P., & Kumar, A. (2015, December). Building enterprise-grade internet of things applications. In *2015 Asia-Pacific Software Engineering Conference (APSEC)*, IEEE, pp. 4–5.
21. Kanawaday, A., & Sane, A. (2017, November). Machine learning for predictive maintenance of industrial machines using iot sensor data. In *2017 8th IEEE International Conference on Software Engineering and Service Science (ICSESS)*, IEEE, pp. 87–90.
22. Meidan, Y., Bohadana, M., Shabtai, A., Guarnizo, J. D., Ochoa, M., Tippenhauer, N. O., & Elovici, Y. (2017, April). ProfilIoT: A machine learning approach for IoT device identification based on network traffic analysis. In *Proceedings of the symposium on applied computing*, ACM, pp. 506–509.
23. Meidan, Y., Bohadana, M., Shabtai, A., Ochoa, M., Tippenhauer, N. O., Guarnizo, J. D., & Elovici, Y. (2017). *Detection of unauthorized iot devices using machine learning techniques.* arXiv preprint arXiv:1709.04647.
24. Ahmed, F. (2017, October). An IoT-big data based machine learning technique for forecasting water requirement in irrigation field. In *International conference on research and practical issues of enterprise information systems*, Springer, pp. 67–77.
25. Patil, S. S., & Thorat, S. A. (2016, August). Early detection of grapes diseases using machine learning and IoT. In *2016 second international conference on Cognitive Computing and Information Processing (CCIP)*, IEEE, pp. 1–5.
26. Pandey, P. S. (2017, July). Machine learning and IoT for prediction and detection of stress. In *2017 17th International Conference on Computational Science and Its Applications (ICCSA)*, IEEE, pp. 1–5.
27. Tallapragada, V. S., Rao, N. A., & Kanapala, S. (2017). EMOMETRIC: An IOT integrated big data analytic system for real time retail customer's emotion tracking and analysis. *International Journal of Computational Intelligence Research, 13*(5), 673–669.
28. Haldorai, A., & Kandaswamy, U. (2018). Cooperative spectrum handovers in cognitive radio networks. In *EAI/Springer innovations in communication and computing* (pp. 47–63). Cham: Springer. https://doi.org/10.1007/978-3-319-91002-4_3.
29. Siryani, J., Tanju, B., & Eveleigh, T. J. (2017). A machine learning decision-support system improves the internet of things' smart meter operations. *IEEE Internet of Things Journal, 4*(4), 1056–1066.
30. Pradana, A. D. I. T. Y. A., Goh, O. S., & Kumar, Y. J. (2018). Intelligent conversational bot for interactive marketing. *Journal of Telecommunication, Electronic and Computer Engineering (JTEC), 10*(1–7), 1–4.
31. Galletta, A., Carnevale, L., Celesti, A., Fazio, M., & Villari, M. (2017). A cloud-based system for improving retention marketing loyalty programs in industry 4.0: A study on big data storage implications. *IEEE Access, 6*, 5485–5492.
32. Fang, S., Da Xu, L., Zhu, Y., Ahati, J., Pei, H., Yan, J., & Liu, Z. (2014). An integrated system for regional environmental monitoring and management based on internet of things. *IEEE Transactions on Industrial Informatics, 10*(2), 1596–1605.
33. Wang, X. V., & Wang, L. (2017). A cloud-based production system for information and service integration: An internet of things case study on waste electronics. *Enterprise Information Systems, 11*(7), 952–968.
34. Rymaszewska, A., Helo, P., & Gunasekaran, A. (2017). IoT powered servitization of manufacturing – An exploratory case study. *International Journal of Production Economics, 192*, 92–105.
35. https://www.finoit.com/blog/enterprise-challenges-in-iot/
36. Heikkila ̈, J. (2002). From supply to demand chain management: Efficiency and customer satisfaction. *Journal of Operations Management, 20*(6), 747–767.
37. Dejonckheere, J., Disney, S. M., Lambrecht, M. R., & Towill, D. R. (2003). Measuring and avoiding the bullwhip effect: A control theoretic approach. *European Journal of Operational Research, 147*(3), 567–590.

38. Carbonneau, R., Laframboise, K., & Vahidov, R. (2008). Application of machine learning techniques for supply chain demand forecasting. *European Journal of Operational Research, 184*(3), 1140–1154. https://doi.org/10.1016/j.ejor.2006.12.004.
39. Bontempi, G., Ben, T. S., & Le Borgne, Y. A. (2013). Machine learning strategies for time series forecasting. In M. A. Aufaure & E. Zimányi (Eds.), *Business intelligence. eBISS 2012* (Lecture notes in business information processing) (Vol. 138). Berlin/Heidelberg: Springer.
40. Chui, M., Lo¨ffler, M., & Roberts, R. (2010). *The internet of things*. McKinsey & Company. Retrieved from http://www.mckinsey.com/insights/high_tech_telecoms_internet/the_internet_of_things
41. Lee, I., & Lee, K. (2015). The internet of things (IoT): Applications, investments, and challenges for enterprises. *Business Horizons, 58*(4), 431–440. https://doi.org/10.1016/j.bushor.2015.03.008.
42. Bose, I., & Mahapatra, R. K. (2001). Business data mining – A machine learning perspective. *Information & Management, 39*(3), 211–225. https://doi.org/10.1016/s0378-7206(01)00091-x.
43. Ampazis, N. (2015). Forecasting demand in supply chain using machine learning algorithms. *International Journal of Artificial Life Research (IJALR), 5*(1), 56–73.
44. Ampazis, N. (2012). A computational intelligence approach to supply chain demand forecasting. In *Machine learning: Concepts, methodologies, tools and applications* (pp. 1551–1565). Hershey: IGI Global.
45. Mori, J., Kajikawa, Y., Kashima, H., & Sakata, I. (2012). Machine learning approach for finding business partners and building reciprocal relationships. *Expert Systems with Applications, 39*(12), 10402–10407. https://doi.org/10.1016/j.eswa.2012.01.202.
46. Cumby, C., Fano, A., Ghani, R., & Krema, M. (2004). Predicting customer shopping lists from point-of-sale purchase data. In *Proceedings of the 2004 ACM SIGKDD international conference on Knowledge Discovery and Data Mining – KDD'04*. https://doi.org/10.1145/1014052.1014098.
47. Buckinx, W., & Van den Poel, D. (2005). Customer base analysis: partial defection of behaviourally loyal clients in a non-contractual FMCG retail setting. European journal of operational research, 164(1), 252–268.

Chapter 6
Enterprise Architecture for IoT: Challenges and Business Trends

A. Haldorai ⓘ, A. Ramu ⓘ, and M. Suriya ⓘ

6.1 Introduction

The industrial architecture is categorized as a discipline used for holistically and proactively initiating industrial responses to disruptive forces. As such, this form of architecture is vital for analyzing and identifying the relevant execution transition to obtain the desired outcomes and visions. This paradigm determines the manner in which enterprises can attain its present and future industrial missions and visions. Industrial architecture includes the management principles, which are behind the present discussions concerning the enterprises strategies and the manner it is presented via the IoTs. The enterprise architecture (EA) was established as an approach to enhance the specification of a logical blueprint, which explains the operation and structure of an enterprise. The purpose of the EA is to determine the manner in which enterprises can effectively attain their present and future missions and visions. EA aligns technology architecture capabilities with business architecture processes and services [1]. Historically, this alignment embraced monolithic structures, including operational systems and data warehouses supported by relational databases and flat files. With today's Internet of Things (IoT) and all the connected devices it involves, solution design offers new opportunities, even though designs can be more challenging. The emerging solution environment has the following characteristics:

A. Haldorai (✉)
Sri Eshwar College of Engineering, Coimbatore, Tamil Nadu, India

A. Ramu
Presidency University, Bangalore, India

M. Suriya
KPR Institute of Engineering and Technology, Coimbatore, Tamil Nadu, India

© Springer Nature Switzerland AG 2020   123
A. Haldorai et al. (eds.), *Business Intelligence for Enterprise Internet of Things*,
EAI/Springer Innovations in Communication and Computing,
https://doi.org/10.1007/978-3-030-44407-5_6

- New services and processes for business commerce, lifestyle, social interaction, culture diversity, and environment management
- A broad range of user platforms such as wearables, smartphones, and a multitude of appliances
- An expanding use of sensors that replicate human sight, hearing, smell, taste, and touch and implement real-time concepts.
- Better timing dependency among and between services and processes using large data volumes and new data analytics to drive the process of decision-making.

This executive update explores the impact of the IoT on traditional business and technology architectures and the role of EA as an effective methodology for developing and implementing IoT strategies. We examine business architecture and how it integrates IoT-driven processes with traditional processes. The IoT is characterized by "things" – many of them small in physical size, which can connect to other things, generating a large network of collaborative IoT and non-IoT things. In this paradigm, business process management (BPM) extends to accommodate innovative workflows that are more functionally robust with "thing-to-thing" linkages. The architects behind the EA are the experts who manage every structure to ensure that IT frameworks are designed based on the present enterprise standards and strategies. The IoTs has proposed a new prototype whereby the global networks of devices and machines are capable of sharing to establish a digitalized innovation in industries. As such, the sector of the IoTs has significantly developed to become the largest sector.

IoTs is large when both viewed as an opportunity for enterprise, including the kind of data it creates. Approximately 90% of data in the whole world has been formed over the past 2 years. Presently, there is an output data of about 2.5 quintillion bytes produced daily. Due to the movement of enterprises from the initial phase of experimentation to complete deployment of the IoTs application, the deluge of data will progressively be termed as plague to the organizations since they focus on processing, capturing, and acting on the big data.

6.2   IoT Trends and Challenges

For the purpose of determining the dependability and reliability of the IoTs, a number of minimal segments of the measures have to be accomplished to attain an interoperability and integration in the sector of IoTs.

Contract decoupling: The IoTs framework includes the heterogeneous networking devices with the disparate communication paradigms [1]. The integration framework has to be competent to effectively deal with contract decoupling.

Scalability: Provided the developing condition of the IoTs and calculation and prediction by [2, 3], an effective incorporation system has to be evolvable and scalable to efficiently support a lot of things linked to the network.

Ease of testing: The integration segment must support the ease of debugging and testing. Moreover, it must promise the support for debugging failures and defects, incorporation testing, system testing, component testing, installation testing, compatibility testing, non-functional testing, security testing, and performance testing.

Ease of development: The IoTs integration system has to promise a form of ease of development for Internet developers. The system should potentially eliminate all the unnecessary complexities to assure an efficient documentation form for both the developers and the non-developers who have to apply the basic programming skills. This enables users to easily understand the inner network systems.

Fault tolerance: The IoTs framework should be resilient and dependable. The smart integration system has to effectively deal with any faults as the IoTs devices can possibly toggle over the online and offline states.

Lightweight implementation: The incorporation frameworks must be lightweight overheads in the deployment and development stages. Moreover, it must be easier to install, activate, uninstall, update, deactivate, and adapt and for versioning [4].

Internal domain operability: The system must be extensible to effectively support the inner domain communication. For instance, in the intelligent vehicle domain, the integration system must promise the support of interaction and communication with the necessary road closures and traffic lights that belong to the intelligent town domains.

6.3 Enterprise IoT – Overview

The enterprise IoTs is the upcoming sector in technology since it will enhance the development of the physical things comprising embedded computer devices to effectively be applied in various business processes. As such, this enables the reduction of manual work while advancing the general organizational efficacy [5]. With the application of the various technological advancements that range from the embedded networking devices with actuators and sensors to the network-based cloud platforms and communication, the enterprise IoTs applications can potentially automate the various enterprise processes. These procedures depend on the contextual data that is produced from the programming devices like vehicles, equipment, and machines. The enterprise IoTs is purposed to be the next advancement in technology, which included the physical things comprising embedded computing networking devices which are used in organizational processing [6]. This process is applicable in reducing manual tasks hence enhancing the overall enterprise efficacy. Based on the combination of a number of technologies that range from the embedded devices with actuators and sensors to the Internet-based communication and the cloud platforms, the industrial IoTs application can effectively automate the organizational processes, which are dependent on contextual data produced by the programming devices like vehicles and machines. Moreover, these enterprise IoTs applications can transfer control instructions to the devices dependent on the relevant enterprise guidelines. The IoTs is a crucial novel networking concept that permits the previously unlinked physical devices to link up with the Internet. In the future, the IoTs will become the Internet itself due to the possibility of availing a lot of trials and opportunities to users [7].

The IoTs is projected to develop the efficiency of enterprises [3] through facilitative novel business models to align various physical processes with digitalized assets on actual-time basis. The big data and cloud data technologies are vital since they support the IoTs in ensuring smart insights and scalability. The main objective of the technology is to formulate an insight that allows a faster, appropriate, and accurate decision-making process in a more client-centric enterprise.

Some of the basic features of every IoTs stack include:

- Loosely coupled: Three critical IoTs stacks have been recognized. However, it is fundamental that every stack is applied separately from the other. It has to be possible to apply an IoTs cloud platform from a single distributor with an IoTs gateway from another supplier and device stacks from a third distributor.
- Modular: Every stack has to permit for the characteristics to be sourced from various distributors [8].
- Platform independent: Every stack has to have its own entire cloud and hardware infrastructure. For instance, the device stacks have to be available on a lot of MCUs and the IoTs cloud platforms, which have to be operated on various cloud PaaS.
- Dependent on available standards: Information transfer between different stacks has to be centered on available standards to accomplish interoperability.
- Defined APIs: Every stack has to contain a defined API, which permits an easy incorporation with the present integration and application of the relevant IoTs solutions [9].

6.3.1   Enterprise IoT Platform: Key Attributes

To take the pulse of enterprise IoT, you need an IoT platform whose key attributes align with the requirements of your enterprise solutions. Your enterprise IoT platform should provide you with the following capabilities:

1. *Managing a large number of devices* –Your enterprise IoT solution requires a flexible device management system that can easily be on-board or off-board and manage a large number of devices in bulk. Plus, it should provide a mechanism to support a variety of contemporary and emerging protocols to deliver the full control of devices, regardless of their make, model, or manufacturer. Finally, your IoT platform should scale to meet the pace and magnitude of growth of connected devices. This kind of platform can store a warehouse of data and manage billions of transactions in real time.
2. *Multi-tenancy and fine-grained access control* – We should emphasize a major difference between creating several customers, a.k.a. tenants, on the platform and creating several users with different permissions within a single tenant.

Tenants are physically separated data spaces with own users, separate application management, and no sharing of data by default. Your enterprise solution needs a natively multi-tenant platform which would allow you to host countless customers running differently branded digital applications without the risk of data leakage. Within a single tenant, you also need fine-grained control over your users and their access rights. For example, not all employees should have access to every device or its data, nor should they have the ability to change the settings. That means such devices should have ways to display what data exactly they are sharing and who they are sharing it with.

3. *Integration with enterprise systems* – Enterprise IoT platform should provide you with an easy way to add new integrations to enterprise systems. By integrating with existing enterprise applications, you can create a master application that would let you streamline your work and get the most value from your data.

4. *Security and privacy at every level of the stack* – Security should be embedded at every level of the platform. Your platform should provide end-to-end data protection ensuring that personal data is not accessible to unauthorized viewers. Most importantly, it should support per-device authentication and authorization to enforce enhanced security. Also, the platform needs to offer a robust and proven security model that will keep your system secure and data safe. Finally, it should be ISO 27001 accredited, which demonstrates that you have taken all the necessary steps to implement internal security practices and protect your business.

5. *Data analysis* – Data analytics should be an integral part of the enterprise connected solution. Why? Through data analysis, raw data is transformed into meaningful information that will help end users to draw key insights to make their decisions move forward.

6.3.2 Enterprise IoT Architecture

The IoT technology stack is given in Fig. 6.1 which contains three tiers: gateways, data center or cloud IoT platform, and sensor devices. According to [10], "a typical IoT solution is characterized by many devices (i.e., things) that may use some form of gateway to communicate through a network to an enterprise back-end server that is running an IoT platform that helps integrate the IoT information into the existing enterprise."

The devices tier concentrates on data collections through the sensors, which can be embedded in various forms of devices. These devices include the mobile computing devices, autonomous appliances, machines, and wearable technologies. These devices have the potential to capture the details concerning the physical environments like light, humidity, chemistry, vibration, and pressure. The standard-centered wireless and wired network paradigms are applicable in the transfer of telemetry information ranging from the cloud via the gateways to the networking devices. The layers of devices form the foundations of the IoTs stacks. The legacy networking devices, which have been available for many years with the intelligent, modern, and

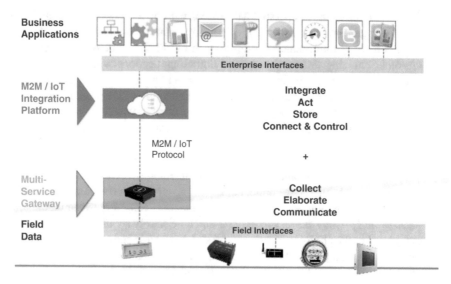

Fig. 6.1 Enterprise IoT architecture model

interlinked devices, are the vital foundation in IoTs. Every device is fundamental for obtaining information from the different sensors, which have the potential to track the crucial parameters. These devices can also be applied on controlling the condition of equipment, for instance, switching off the faulty machines or rectifying the RPM of a revolving gear. These device layers have the capacity to assure the final mile connection to the remote equipment. It therefore presents the present status of network devices alongside the capability to remotely control these devices.

The gateways, which are also known as the control tiers, are known as the intermediaries that potentially facilitate the transfer of information, offload processing actions, and drive functions. Due to the fact that some sensors produce a lot of data points in every second, these gateways assure a place for pre-processing of data locally before being sent to the relevant cloud tiers. Whenever information has been aggregated in the gateways, it is summarized and strategically evaluated. Limiting the amount of information can have a significant implication on the networking transmission costs, mostly on mobile networks. Moreover, this permits a crucial organization rule, which is applicable depending on the amount of data streaming in. The control tier is considered bidirectional. This can possibly provide control data to the relevant devices to facilitate configuration transitions. During the same moment, it can possibly respond to information-tier command-and-control requests, like the security requests used in the process of authentication.

The low-power and legacy devices cannot be used to directly register or communicate data to the various IoTs platforms. In this process, the gateways are considered in the entire networking picture. This is sometimes referred to as the edge devices, whereby the gateways are used as a proxy to the various networking devices. These gateways are accepted to route the commands that are transferred from the back end to the relevant networking devices. An edge device or the locally available gateways present a significant market chance. The networking and hardware vendors are focusing to capture their own share of the market through the process of augmenting their switches, routers, edge devices, and firewalls meant to double up the various IoT gateways. The inclusivity of the novel IoT devices, edge devices, and legacy machines forms a crucial device layer of the IoTs stacks.

The cloud tier, data center, and the IoTs platform are obliged to undertake a large-scale computation of information to produce the relevant insight, which is vital in the enterprise. It is therefore relevant in offering a back-end enterprise analytics to fully execute a complex event processing like the process of analyzing information that is adaptable in the business world with rules that are inclined to historical trends. Moreover, this process includes the dissemination of organizational guidelines downstream. The process requires scalability both horizontally and vertically to effectively support an increasing number of linked devices. These are meant to address the various IoTs problems. The vital functions of the IoTs information center and cloud platform include the messaging routing and connectivity, data storage, event analysis, device management, and application enablement and integration [11]. The functioning capacity of the industrial IoTs depends on the software platforms, which manage the status of devices, analyze them, store them, and present the correct insights to enhance the process of decision-making. Moreover, it acts as the middleware, which is meant to orchestrate the complete flow of work. Provided the attributes of the cloud like reliability, scale, and elasticity, it is currently becoming a crucial deployment environment of the IoTs platform.

The device layer denotes the cloud gateways, which are responsible for authorizing and authenticating devices meant to manage the flow of works. Moreover, it ensures that secure communication is maintained between the various devices and the centralized command centers. These gateways are purposeful in dealing with a lot of protocols and the information formats. The heterogeneous device and locally available gateways with disparate protocols are used in the process of registering the cloud gateways. For instance, the locally available gateways and the devices can possibly communicate with the cloud gateways via the SOAP, REST, AMQP, XMPP, CoAP, MQTT, and WebSocket. Despite the inbound protocols, the cloud gateways are significant for assuring a consistent view of devices to the remaining components. An average enterprise IoTs deployment is responsible for thousands of devices and sensors that have to be deployed in a lot of sites. Every device requires registration and maintenance in the centralized repository that serves as the authoritative inventory that acts in place of the present condition of deployment. The device registry represents a centralized inventory where each device is registered into a system. Every device alongside the metadata is available in the individual registry. Any component in a platform can query the network device registry to evaluate the present status of a device alongside its capability.

6.4   IoT Business Model

Progressively, all forms of businesses are purposed to replace their one-off and discrete sales frameworks with the present subscription-centered framework, which links up products and services within a framework for a long-term relation. These subscription models are effective and relevant to both the customers and businesses. The business is designated to more recurrent and consistent streams of revenues. Clients are no longer subjected to massive capital investments but they are being offered opportunities of scaling their capacities meant to access superior services and support. In the current years, advance cloud computing, technologies, and connectivity have made the process possible since the advent of the "as-a-service" enterprise model. This process began with the software-as-a-service (SaaS) before becoming adjusted software. The IoTs has assured a powerful novel enterprise framework, which is considered as a shot in the arm. Moreover, the low-cost, interlinked IoTs sensors and the monitored devices, including the sold products, enhance the kind of services produced in an enterprise. The actual-time and historical information generated by the IoTs devices permits users to exercise preventive maintenance. For instance, the automated alerting and the predictive maintenance have become possible in the modern age. These developments are useful in enhancing the technological investment connected to the IoTs [12].

6.4.1   Types of IoT Architecture Models

Device-to-device: "IoT devices within the same network that generally connect using wireless PAN protocols, such as Bluetooth and ZigBee, are device-to-device architectures," the report illustrates.

Device-to-cloud: In such architectures, "IoT devices connect directly to the cloud, typically using a long range communications network, such as cellular. For example, IoT-enabled vehicle monitoring devices (such as those provided by car insurance companies to drivers) collect data on the vehicle, such as distances and speeds driven, and acceleration and braking rates. These data are then transmitted to the cloud, analyzed in the cloud, and used by insurance companies to create tailored insurance rates based on the driving data."

Device-to-gateway: "Device-to-gateway architectures transfer information from sensors to the cloud via a gateway device. The gateway collects the data and then communicates the data to the cloud through additional network connectivity, such as Wi-Fi or cellular connection."

Cloud-to-cloud: "Cloud-to-cloud architecture, also known as back-end data sharing, enables third parties to access uploaded data from IoT devices. For example, smart buildings receiving data from smart thermostats and smart light bulbs can send the data to a cloud via Wi-Fi. The collected data are then aggregated in cloud 1, which may be owned by the building as the conduit, a user can set a smart

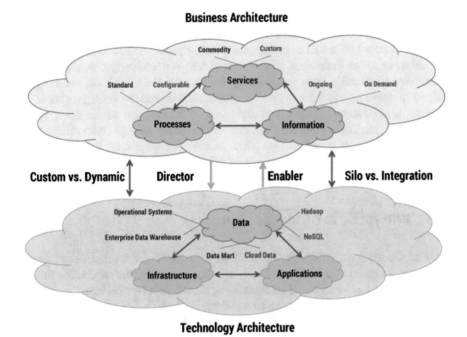

Fig. 6.2 Types of IoT architecture models

thermostat to activate when the user's car approaches their house (the conditional trigger in this example being the location sensor within the car)."

IOT-enabled enterprise architecture given in the figure is a blueprint for the deployment of IT resources that support business services and processes. It is a high-level view of six fundamental domains: service, process, information, application, data, and infrastructure.

EA aligns new technology capabilities with innovative business architecture processes and services. In Fig. 6.2, new architecture elements associated with the IoT are accented in red. In the business architecture with IoT, commodity services with standard processes and operational data are augmented with custom services delivered through configurable processes. These processes provide on-demand information tailored to user preferences or customized processes delivering unique services associated with multiple, diverse connected devices. Due to larger consumer markets, more configurable processes will be required to meet broader customer preferences [13].

In the emerging EA [14] in which the IoT is a force, the architecture is more diverse and complicated. New data platforms, including Hadoop, NoSQL, and cloud data, will be used to house huge volumes of data. The IoT will enable people in their daily lives to access custom services implemented using configurable processes. For example, a smart refrigerator user can set food expiration policies and rely upon automated leading signals before actual expiration dates are realized.

A configurable process minimizes food loss and reduces consumer grocery cost. In another example, a food manufacturer can signal consumers if a product has been found to pose public health risks (e.g., the possible presence of *E. coli*) to minimize consumer exposure to this risk. With real-time design, a food producer might search purchased smart refrigerator data, identify the location of tainted food, and send an urgent message to affected consumers, which can be broadcast among a cluster of product users, increasing the speed and reach of the message within the targeted customer sector. These innovative workflows are enabled through smart refrigerator integration with applications and data located across selected infrastructure platforms.

6.5   Enterprise IoT Monetization

One of the most vital objectives of the IoTs is to transform the mindset of doing things in an enterprise. This mindset is relevant in the process of determining if users can use a product-based monetizing approach or a service-based approach. CSPs, ISVs, OEMs, and other relevant stakeholders are invited to contribute their intellectual properties in advancing a significant interlinked ecosystem. This process necessitates monetization of a framework meant to permit all the contributors to effectively leverage the novel IoTs organizational models to assure the agility to deploy novel application to accomplish quicker ROI. The application enablement platforms such as SensorLogic are vital for bridging this necessity with a competent device pre-built and on-boarding to the IoTs services.

6.5.1   Enable Flexible Monetization Models

The IoT is developing a novel enterprise framework, for instance, the PaaS (product-as-a-service), whereby the OEM provides the device, but in the instance of charging the client upfront, the OEM permits them to pay via flexible frameworks (monthly, pay-per-use or metered, etc.). The Sentinel Software Monetization resolutions permit the CSP and OEM to effectively apply the flexible monetization frameworks via tested technologies.

6.5.2   License and Entitlement Management

The objective of software is gradually developing and will continue progressing since there are a lot of interlinked applications which have emerged in the recent years. ISVs have, for a long time, handled various problems to enhance the monetization of their smart properties, i.e., software. This involves challenges such as reverse engineering and piracy. Whereas the software industries have increased, the incumbent and emerging device producers are still new to their pending concerns. To effectively monetize the intellectual properties, there should be effective enforcement of the licensing policies for all the devices, which must also be broken down to the individual features in the devices. The Sentinel Embedded and Cloud Service Monetization resolutions permit the OEM and cloud service vendors to effectively enforce the licensing frameworks.

6.5.3 Software Upgrades

Due to the rapid development of the IoTs services, the OEM will necessitate a robust infrastructure for the purpose of remote upgrading of features and software on the devices. However, these upgrades necessitate advanced levels of reliability and security to assure that the devices in the fields cannot be tampered with making them safe. The Sentinel Embedded resolutions assure a system for secure downloading of the software upgrades, which make sure that the features can be deployed for the purpose of generating new streams of revenues from the interlinked devices.

6.5.4 Device Management

As a result of proliferation of different endpoints, the difficulty of different device protocols has gradually enhanced. Ranging from the standard-based to the proprietary implementation, the device integration into the IoTs environment is considered as a time-consuming process. To assist devices to communicate with the business or cloud resources, the SensorLogic platform is relevant as an orchestrator. Based on the application of the device translators, the device on-boarding is considered easier, which reduces the set-up process of the devices in the network.

6.5.5 IoT Application Development

The SensorLogic platform promises a system that can be used for the rapid development of novel applications for IoTs use cases. This system involves (but is not limited to) a wide range of web services for the purpose of administration, authentication, alarm notification, and location identification. Moreover, it assures a pre-integration of a third-party service such as connectivity and geo-fencing resolutions. The pre-developed building blocks are used by IoTs developers to enhance the formulation of new use cases, in addition to the present blocks. Developer kits, such as the Cinterion Concept Board, and application platforms, such as Java, are relevant for prototyping and designing. For the client, this represents a quicker timeframe of deployment and ROI for the purpose of IoTs application.

6.6   Case Study

6.6.1   *Waste Management*

M1 Limited (M1), one of Singapore's leading full-service communications providers, is working with OTTO Waste Systems Singapore Pte. Ltd. (OTTO), to provide
a litter bin management system to the National Environment Agency (NEA). The
new system is designed to enable the NEA to leverage technology to better manage
the deployment of litter bins, as well as to optimize cleaning resources. The NEA
has been exploring how data can be used to enhance the effectiveness and efficiency
of public cleaning. The new system utilizes IoT sensors fitted within litter bins to
track how full they are, so that the cleaning crew can be notified when they need to
empty these litter bins. The NEA can also monitor the usage of litter bins to gauge
if there are adequate bins in a particular area to serve the public. OTTO aims to
deploy up to 500 of these smart bins during the first quarter of 2019. M1 says the
reliable and secure city-wide coverage provided by its NB-IoT, together with its
support for industry standards, makes the technology well suited for large-scale
smart city applications, such as the proposed litter bin management system. Smart
city solutions can also benefit from NB-IoT's power efficiency, which makes it viable to use batteries in connected devices, thus reducing infrastructure and maintenance costs. M1 developed the litter bin monitoring solution together with OTTO,
who supplies the litter bin receptacles, and SmartCity, who provides the centralized
management system. "The collaboration with M1, using their NB-IoT network for
smart waste management, allows our customers to enjoy easy access to useful real-
time data for smarter planning and resourcing on waste management nationwide,"
says Christopher Lopez, Managing Director of OTTO Waste Systems. "We also see
the potential of such implementations to help consumers have a greener environment to live in." "Extensive research and development were carried out to produce
the hardware and the methodology of installation to maximise the accuracy of the
measurement in waste level and pollution in the environment," adds P. Renganathan,
Director of SmartCity. "Through the strategic cooperation with M1, we will help
companies to achieve greater cost savings and reach higher productivity."

6.6.2   *Smart Cold Chain – Tracking Temperature During Transit*

In Thailand, mobile operator AIS has developed a mobile IoT-based solution for
monitoring the temperature of perishable goods during transportation as depicted in
Fig. 6.3. Fresh food, frozen food, medicine, and some other goods need to be kept
at a constant temperature during distribution to ensure that they don't decay and that
they reach end customers in a pristine condition. Connected "cold chain" solutions
can be used to monitor the temperature of a cold storage container during transit and
maintain the quality of goods, reducing the number of claims from customers that
goods have decayed or been damaged during transportation.

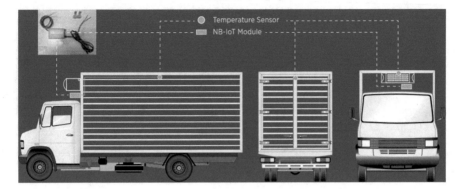

Fig. 6.3 Smart cold chain model

These solutions can be configured to send a notification to the supplier if the temperature rises beyond a specific threshold. To meet the demand for a low-cost solution that can be installed easily without impacting logistics companies' existing systems, AIS is using NB-IoT to connect on-board thermometers to its IoT platform, which can be used to record, analyze, and display the resulting temperature data. AIS says the compact battery-powered thermometer is cost-effective enough to be deployed at scale, while its small size and independent power supply mean it can be quickly and easily installed or moved to another location. The device can measure the temperature between −50 and 20 degree Celsius. It can be configured to transfer temperature data to AIS IoT platform every x interval such as every 3 min and alert when the temperature changes by more than x degree Celsius such as one degree Celsius. AIS says it is also using NB-IoT to monitor the electrical supply of the cooling system, allowing it to ensure there is sufficient power to cool the goods being transported. If the power supply is not working properly, the system is designed to relay the relevant data to the AIS IoT platform, thereby allowing the logistics company to proactively resolve the issues before any serious damage is caused. "By installing the temperature-measuring devices in cold chain logistic systems, the quality of perishable goods can be assured – Mobile IoT connectivity can be used to notify the operating parties when issues arise and take necessary actions to prevent any damage to the goods," explains Asnee Wipatawate, Head of Enterprise Product and International Service of AIS. "The quality of IoT solutions becomes critical to mitigate this problem and therefore yield competitive advantages."

6.6.3 Smart Security

The remote access management and monitoring of high-value, distributed infrastructure assets is a historically difficult, labor-intensive, disconnected, and non-scaling burden. Singtel's vision of a frictionless, smart, highly scalable perimeter access control solution begins today with the planned commercial launch of igloohome's connected digital lock system as shown in Fig. 6.4. Singtel and

Fig. 6.4 Smart security lock

igloohome (a Singtel-funded start-up) are excited to announce the upcoming commercial launch of their connected perimeter access solution, based on igloohome's connected lock technology.

The resolutions assure an actual-time management, scalable remote control, and monitoring of the distributed infrastructure perimeter accessibility. igloohome works with leading property developers throughout Asia, including but not limited to Sansiri (Thailand), Capitaland (Vietnam), and Mitsubishi (Japan). Matthew Ng, VP of Product of igloohome says: "We have adopted carrier-grade IoT network technologies like LTE-M and NB-IoT as they are increasingly prevalent among global operator IoT solutions deployments. In Singapore, we rely on Singtel's NB-IoT cellular network because of its wide coverage and high availability. These high-quality public IoT networks give us faster time-to-market, and obviate the need to deploy our own private network, or implement discrete connectivity hubs/ gateways. LTE-M and NB-IoT are very power efficient, making our battery-powered smart locks more appealing to end users." igloohome is a worldwide partner of Airbnb, Booking.com, and Agoda, works with over 50 distributors, and ships to more than 90 countries. A global operator-deployed standard like NB-IoT is an essential element in support of igloohome's global business.

Rahul Mehta, IoT Product Lead of igloohome comments: "We have extensively tested our NB-IoT locks across many countries in Asia, and they have performed well among all deployment scenarios – from deep indoor, to outdoor, and even remote locations. We are excited to meet the global market demand for our connected smart locks solution, a task that's simplified by a global IoT network standard like NB-IoT." igloohome first proved its technology in the vacation rental space, partnering with Airbnb to simplify host-controlled guest access without the need for a physical key exchange. The solution further enhanced host peace of mind with on-demand, detailed visibility of guest-specific room, site, or location access. igloohome then broadened its offering to address the needs of different categories of home and property owners, addressing the operational and security limitations of physical keys and enabling use cases like time-sensitive, remote-monitored, and controlled access for delivery and trade services and access expiration for former tenants.

6.7 Summary

Enterprise architecture is an important methodology for improving and cultivating the business value of IT. The emerging capabilities of the IoT are valuable processing of data and analytical contributions to new enterprise services and processes. Consequently, the business value of the IoT can be realized and validated when an EA framework is applied to IoT strategy and projects. When an EA framework for IoT is constructed, it is not isolated from non-IoT components. Architectures should integrate non-IoT and IoT into a unified blueprint. This framework is composed of reference models, standards, and principles that encompass the expanded functionality of new non-IoT and IoT products and services. Today, new data technologies and flexible platform configurations can be implemented to manage this data. From a market perspective, user-enabled products are a major growth sector targeted by many companies. Popular user-enabled products, including smartphones and wearables, are in demand in broad consumer markets. User-enabled products are increasingly powered by the IoT.

References

1. Shroff, G. (n.d.). Enterprise architecture: Role and evolution. *Enterprise Cloud Computing*, 39–48. https://doi.org/10.1017/cbo9780511778476.006.
2. Kale, V. (2019). Internet of things (IoT) computing. In *Digital transformation of enterprise architecture* (pp. 413–435). Milton: CRC Press. https://doi.org/10.1201/9781351029148-22.
3. Tambo, T. (2017). Enterprise architecture beyond the enterprise – Extended enterprise architecture revisited. In *Proceedings of the 19th international conference on enterprise information systems*. https://doi.org/10.5220/0006277103810390.
4. Anandakumar, H., & Umamaheswari, K. (2017, March). Supervised machine learning techniques in cognitive radio networks during cooperative spectrum handovers. *Cluster Computing, 20*(2), 1505–1515.
5. Anandakumar, H., & Umamaheswari, K. (2018, October). A bio-inspired swarm intelligence technique for social aware cognitive radio handovers. *Computers & Electrical Engineering, 71*, 925–937. https://doi.org/10.1016/j.compeleceng.2017.09.016.
6. Arulmurugan, R., & Anandakumar, H. (2018). Early detection of lung cancer using wavelet feature descriptor and feed forward back propagation neural networks classifier. In *Lecture notes in computational vision and biomechanics* (pp. 103–110). Cham: Springer. https://doi.org/10.1007/978-3-319-71767-8_9.
7. Haldorai, A. R., & Murugan, S. Social aware cognitive radio networks. In *Social network analytics for contemporary business organizations* (pp. 188–202). https://doi.org/10.4018/978-1-5225-5097-6.ch010.
8. Arulmurugan, R., & Anandakumar, H. (2018). Region-based seed point cell segmentation and detection for biomedical image analysis. *International Journal of Biomedical Engineering and Technology, 27*(4), 273.
9. Suganya, M.,& Anandakumar, H.(2013,December). Handover based spectrum allocation in cognitive radio networks. *2013 International Conference on Green Computing, Communication and Conservation of Energy (ICGCE)*. https://doi.org/10.1109/icgce.2013.6823431, https://doi.org/10.4018/978-1-5225-5246-8.ch012.

10. Implementation: Developing Enterprise Architecture. (n.d.). *From enterprise architecture to IT governance*, 167–194. https://doi.org/10.1007/978-3-8348-9011-5_7.
11. Rana, M. M., & Bo, R. (2019). IoT-based cyber-physical communication architecture: Challenges and research directions. *IET Cyber-Physical Systems: Theory & Applications*. https://doi.org/10.1049/iet-cps.2019.0028.
12. Al-Qaseemi, S. A., Almulhim, H. A., Almulhim, M. F., & Chaudhry, S. R. (2016). IoT architecture challenges and issues: Lack of standardization. *2016 Future Technologies Conference (FTC)*. https://doi.org/10.1109/ftc.2016.7821686.
13. Planning: Creating Enterprise Architecture. (n.d.). *From enterprise architecture to IT governance*, 153–166. https://doi.org/10.1007/978-3-8348-9011-5_6.
14. Armour, F., Kaisler, S., & Bitner, J. (2007). Enterprise architecture: Challenges and implementations. *2007 40th annual Hawaii International Conference on System Sciences (HICSS'07)*. https://doi.org/10.1109/hicss.2007.211.

Chapter 7
Semi-Supervised Machine Learning Algorithm for Predicting Diabetes Using Big Data Analytics

Senthilkumar Subramaniyan, R. Regan, Thiyagarajan Perumal, and K. Venkatachalam

7.1 Introduction

This chapter introduces the concept of Big Data for predicting complications of diabetes mellitus. The discussion includes descriptive and predictive investigations, knowledge discovery and information disclosure, cloud processing, Hadoop MapReduce, and machine learning strategies.

7.1.1 Machine Learning and Knowledge Discovery

1. **Big Data**

Big Data uses an assortment of components: social systems administration, mobile computing, data analytics, and the cloud, which are collectively known as SMAC. Modern research areas require huge amounts of information that are difficult to store and process quickly in a reasonable amount of time using customary approaches. Thus, such information requires an innovative approach, such as Big Data [1]: "Big data alludes to the instruments, procedures, and strategies enabling an association to make, control, and oversee gigantic informational indexes and storage facilities" [2].

S. Subramaniyan · T. Perumal
Dept of CSE, University College of Engineering, Pattukkottai, Rajamadam, India

R. Regan
Computer Science and Engineering, University College of Engineering, Villupuram, India

K. Venkatachalam (✉)
School of Computer Science and Engineering, IT Bhopal University, Bhopal, India

© Springer Nature Switzerland AG 2020   139
A. Haldorai et al. (eds.), *Business Intelligence for Enterprise Internet of Things*,
EAI/Springer Innovations in Communication and Computing,
https://doi.org/10.1007/978-3-030-44407-5_7

2. Big Data in Healthcare

The rapid growth of information in healthcare applications is related to the digitization of patient information, installed frameworks in social insurance, and the need for accessible and efficient mobile systems. Breaking down and deciphering the enormous and complex datasets in healthcare is a challenge for specialists and is known as a Big Data issue.

Big Data management is a significant job in healthcare. To a great extent, Big Data analytics enables patients Well-Being of mind, framework and legitimate management. Big Data methodology considers issues related to the 4 Vs: volume, velocity, variety, and veracity.

3. Diabetes Mellitus

Diabetes mellitus (DM) is a metabolic issue with abnormal insulin regulation. Insulin insufficiency results in higher blood glucose levels and impaired absorption of sugar, fat, and proteins.

DM is a common endocrine issue, affecting approximately 200 million individuals worldwide. The incidence of DM is expected to increase drastically in the coming years. DM can be generally classified as one of two types: type 1 diabetes (T1D) and type 2 diabetes (T2D), based on etiology. T2D is the most common type, chiefly characterized by insulin resistance. Fundamental causes of T2D include lifestyle, physical activity, diet, and genetics. T1D is believed to result from the autoimmunological devastation of Langerhans islets facilitating pancreatic-β cells. T1D accounts for 10% of diabetes cases around the world. Other types of diabetes can be characterized by their insulin emission profiles and include gestational diabetes, endocrinopathies, maturity onset diabetes of the young (MODY), neonatal diabetes, and mitochondrial diabetes. The main complications of DM include polyurea, polydipsia, and critical weight loss. Diagnosis relies upon blood glucose levels (fasting plasma glucose = 7.0 mmol/L).

7.2   Related Work

DM has been unequivocally associated with a few complications, mostly because of the constant hyperglycemia. DM has a broad scope of different pathophysiologies. Many well-known complications have been identified, including vascular issues, diabetic nephropathy, retinopathy, neuropathy, diabetic coma, and cardiovascular disease. Because of the great morbidity and mortality associated with DM, prevention and management have attracted considerable attention. Insulin is a fundamental treatment for T1D, but is only given in specific instances of T2D, such as when hyperglycemia cannot be controlled through diet, weight loss, exercise, and oral medication. The targets of medication aim to

(a) Spare a patient's life and mitigate complications.
(b) Prevent complications and thus extend life span.

The most widely recognized anti-diabetic medications include sulfonylurea, metformin, alpha-glucosidase inhibitors, peptide analogs, and non-sulfonylurea secretagogues. Most of these medications have various side effects. Furthermore, insulin treatment is associated with weight gain and hypoglycemic episodes. Hence, antidiabetic medications and treatments are of extraordinary importance and simultaneously a clincial challenge.

1. Big Data Analytics

An immense amount of information is created each day by specialists, treatment plans, medical reports, imaging, and body sensor devices, such as IoT gadgets, which result in Big Data. Creating Big Data is certainly not a significant challenge for healthcare experts, but obtaining helpful information from Big Data is. Investigations are required to contribute Big Data to medical services. Accessible medical information in a Big Data investigation enable health care providers to provide efficient results, from which patients can be prescribed the best medications or treatments. Currently, Cloud-based Big Data analytic environment is needed to analyze semi structured, structured and unstructured Big Data from healthcare sector.

2. Descriptive/Predictive Analytics

Insightful investigations are fundamental to predict disease in a patient based on the patient's health parameters. Using machine learning, such as support vector machine (SVM) calculations with data frameworks such as Hadoop, it is conceivable to execute a huge volume of data sets effectively and quickly to predict the likelihood of diabetes based on a patient's lifestyle. Along these lines, a Hadoop distribution in a conveyed condition is the most recent innovation.

3. Hadoop MapReduce

A lot of processing capacity is needed to separate helpful information in Big Data. Distribution and/or circulation enable us to take quick calculations by utilizing the numerous CPUs or CPU centers to run a few calculations simultaneously. Although parallel and disseminated registering are not identical, they fill a similar principal need of separating huge problems into smaller ones that can be analyzed simultaneously. In distribution or parallelization, parameters such as execution speed, memory, and simultaneousness are significant elements. The memory limit of a few machines can be used for distribution and/or parallel or circulated handling; however, this can be overcome by using a single device to process the enormous data collections.

4. Machine Learning Techniques

In computer science, artificial intelligence (AI), sometimes called machine intelligence, is intelligence demonstrated by machines, in contrast to the natural intelligence displayed by humans and animals. Machine learning technique takes over the activity involving automated learning from data set. It initially learns the knowledge and applies this knowledge to distribute for predictions. A wide area of artificial intelligence enabled devices are implemented through information.

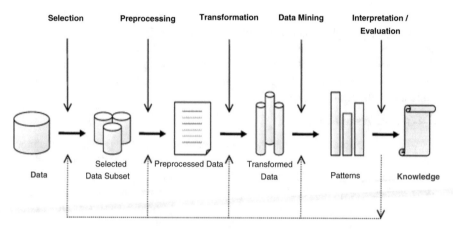

Fig. 7.1 The general flow of a KDD process

The general meaning of machine learning was denoted by Mitchel:

A computer program is said to learn from experience Ex with respect to some class of tasks Ta and performance measure Pe, if its performance at tasks in Ta, as measured by Pe, improves with experience Ex [3].

Knowledge discovery in databases (KDD) is an area associating theories, tools, and techniques to create data and mine useful patterns or knowledge. The many steps of KDD (gathering, feature selection, transformation, knowledge discovery, pattern evaluation and creation) are shown in Fig. 7.1. A full overview of KDD was provided by Fayyad et al., who defined it as follows:

KDD is the nontrivial process identifying valid, novel, potentially useful, and ultimately understandable patterns in data.

Types of Machine Learning Tasks
Machine learning [4] tasks can be categorized into three main types, as shown in Fig. 7.2:

1. Supervised learning – The system can learn the data by labeling 30% as training data,
2. Unsupervised learning – The system can adapt by unlabeling the data
3. Reinforcement learning – The system can communicate with static/dynamic environments.

1. Supervised Learning

In supervised learning, a framework can "learn" things inductively with a capacity known as target work, which is represented as an outflow of an output model depicting the data and information. A target work can be utilized to estimate an input variable, also known as a subordinate instance/variable or a yield instance/variable, depicted from autonomous factors, information factors/attributes, or characteristics. The arrangement for conceivable information estimations of the

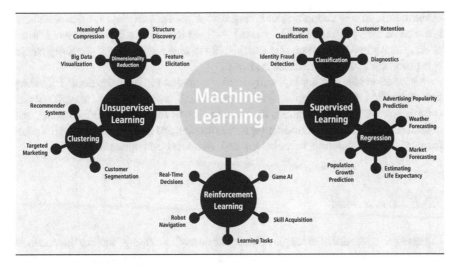

Fig. 7.2 Machine learning tasks

capacity, such as its area, are known as occurrences. Each case can be portrayed with many of the qualities (traits/characteristics). The subset of everything being equal; in which, it provides each and every variable plays an important role. The implementation of data was denoted as a models. Therefore, to derive an objective capacity, learning framework, or preparation subset, theory/elective capacities and known theories, signified by a hypothesis, must be considered. Supervised learning consists of two types of learning methods:

- Grouping
- Relapse

Arrangement techniques attempt to anticipate particular methods, such as blood collections, whereas relapse techniques anticipate numerical qualities. The most commonly used supervised systems are decision trees (DT), rule learning, and instance-based learning (IBL), such as k-nearest neighbors (k-NN), genetic algorithms (GA), artificial neural networks (ANN), and support vector machines (SVM).

2. **Unsupervised Learning**

In unsupervised learning, the machine is used to provide the hidden variable and structure of the input data/associations among the different variables. Therefore, the training data consists of variables without the target labels.

7.3   Association Rule Learning

Association rule mining denotes an interrelationship between data items that is used in many techniques. It is receiving more attention in research on databases [5, 6]. This approach was proposed in the 1990s by Rakesh Agrawal in a market bin examination. In this shopping/marketing basket model, association rules were represented as the structure of $\{A1, ..., An\} \rightarrow B$, which implies that when you find all of A1, ..., An in a truck, then it is conceivable to also find B. The Apriori algorithm was proposed in 1994 by Rakesh Agrawal.

Association rule mining is mainly used for market bin investigations, including bioinformatics. Another application in science and bioinformatics incorporates an organic succession investigation, the examination of high-quality information, and others. A careful survey of finding regular examples and association rules originates from organic information, calculations and unsupervised applications.

7.3.1 Clustering

Clusters are educational examples that occur through grouping, such as the partition of an entire dataset into groups of information [7]. The goal is to gather similar examples for comparison under a particular circumstance.

3. **Reinforcement Learning**

Reinforcement learning is a type of machine learning used to group methods where the framework covers to learn and act through direct contact with nature for enhancing the general ides for execution. Reinforcement learning is predominantly applicable to self-ruling frameworks, because of the freedom to connect its conditions [8].

7.4 Diabetic Mellitus and Machine Learning

The next section describes the significant processes for DM [9].

7.4.1 Biomarker Identification and Prediction of DM

An enormous number of components are regarded as significant in DM. Obesity is considered to be significant hazard impact, particularly in T2D, providing a very strong connection among the types of DM. The diagnosis of DM can be aided by a few tests. In the early stages, the findings in both T1D and T2D can affect the following:

(a) Slowing progression of the disease.
(b) Directed choice of medication.
(c) Preventing future complications.
(d) Identifying related problems.

Biomarkers (BM) (e.g., organic atoms) are quantifiable markers of a specific condition associated with healthy and unhealthy states. Normally, BMs are described as follows.

(a) Found in bodily fluids (blood, saliva, or urine).
(b) Experienced data and information was decided based on the autonomous etio-pathogenic robotic pathway.
(c) Used to screen for disease status and response to medical treatment.

Biomarkers will be the immediate conclusion of the use of infection itself with different inconveniences. Current advances, including metabolomics, proteomics, and heredity, will improve the use of BMs. In cases of DM, BMs can indicate hyper-glycemia and other relevant issues in diabetes.

The following sections discuss diagnostic and predictive markers, as well as disease prediction.

7.4.2 Diagnostic and Predictive Markers

The primary class manages biomarker disclosure, which is a task for the most part that performed through component choice methods. Following an element determination step, an order calculation is used to evaluate the expected precision of the chosen characteristics [10–16].

Primarily, setup techniques are utilized in biomarker assessment. A medical dataset containing 803 prediabetic females with 55 characteristics was used for a few basic element determination calculations to predict DcMs. It presumed that the better presentation that has been accomplished with wrapper strategies. In addition, among the channel strategies utilized, balanced vulnerability showed the best outcomes. It consolidated electromagnetic (E-M) calculations of the closest classifier and an inverse sign test (IST). As a totally unique methodology to manage characteristics in a medical dataset for diabetes, cluster-based (progressive grouping) includes an extraction system using data on disease symptoms. Their approach delivered groups to be used as characteristics for disease severity and patient readmission hazard forecasts.

7.5 Prediction of DM

The second class manages disease expectation and analysis. Various calculations and methodologies have been applied, such as conventional AI calculations, outfit learning methods, and association rule methods, to accomplish the order exactness. In addition to the previously mentioned technique, Calisir and Dogantekin implemented LDA–MWSVM, a framework used to diagnosis diabetes. This framework includes feature extraction techniques, such as Linear Discriminant Analysis (LDA), followed by a Morlet Wavelet Support Vector Machine (MWSVM) classifier.

High-dimensional datasets may be used. For example, one dataset consisted of 4.1 million people and 42,000 factors from regulatory cases, pharmacy records, social insurance use, and laboratory results for the years from 2005 to 2009. This was used to assemble prescient models (in light of calculated relapses) for various onsets of T2D. Various learning calculations appear to be viable methods for improving arrangement exactness. The methodologies have additionally been used in DM prediction.

7.5.1   Diabetic Complications

Complications from elevated glucose levels are an extraordinary field. The negative effects of hyperglycemia can be classified as follows:

(a) Macrovascular complications, such as coronary supply route infection, fringe blood vessel malady, and stroke.
(b) Microvascular complications, including diabetic neuropathy, nephropathy, and retinopathy.

Hyperglycemia is considered to be a fundamental contributor to morbidity and mortality in T1D and T2D. Many studies have demonstrated an association between glycemia and diabetic microvascular complications in both T1D and T2D. Diabetic complications can likewise be grouped by their severity and onset.

Additionally, both insulin resistance and hyperglycemia are involved in the pathogenesis of diabetic dyslipidemia. This is significant because DM affects individuals with well-controlled blood glucose levels. Many of these instances have been considered in AI and information mining applications [17–24].

Huang et al. [19] used a decision tree-based expectation device that joined both genetic and clinical characteristics to identify diabetic nephropathy in individuals with T2D. Leung et al. [20] examined a few AI strategies that incorporated halfway least squares relapse, characterization and relapse trees, the C5.0 Decision Tree, random forest, innocent Bayes, neural systems, and bolster vector machines. The dataset was composed of both genetic (single nucleotide polymorphisms) and clinical information. Age, time of diagnosis, systolic blood pressure, and hereditary polymorphisms of uteroglobin and lipid digestion emerged as the most useful indicators.

DuBrava et al. [21] used random forest to select specific characteristics for predicting diabetic peripheral neuropathy. In order of importance, the characteristics selected were the Charlson Comorbidity Index score (100%), age (37.11%), number of pre-record methodology and administrations (29.701%), number of pre-file outpatient treatments (24.223%), number of pre-list outpatient visits (18.302%), number of pre-file research center visits (16.912%), number of pre-list outpatient office visits (12.12%), number of inpatient treatments (5.913%), and number of torment related medicine remedies (4.404%). The general precision of the model arrived at 89.01%. The Diabetes Preprocessing Research Activity (DiScRi) database was used to predict cardiovascular autonomic neuropathy (CAN).

Staniery et al., [22] used choice trees and ideal choice way discoverer (ODPF) to locate the ideal succession of tests to diagnose CAN. Abawajy et al. used relapse and meta-relapse, in combination with the Ewing equation, to distinguish the classesofn CAN, while conquering the issue of missing information. Cardiac abnormalities are common diabetic complications. They are considered to be a critical connection between diabetes, coronary disease, and stroke.

Hypoglycemia (low glucose levels) mainly occurs from anti-diabetic medications and has an extraordinary effect on individuals with DM. AI strategies such as random forest, bolster vector machines (SVM), k-closest neighbor, and innocent Bayes were used by Wang et al., [24] to predict hypoglycemia in individuals with T2D. Bolster vector relapse was used by Georga et al. for a similar investigation. Furthermore, effectively distributed calculations were used by Jensen in a similar study.

Diabetic retinopathy (DR) is an eye disorder that occurs in individuals with T1D or T2D. The longer a patient has diabetes, the higher is the risk of acquiring this particular pathophysiological condition. D_R is a rule shows early warning sign & it is portrayed with significant diabetic confusion. DR has two primary stages: nonproliferative (NPDR) and proliferative (PDR) Information mining and AI have been used to predict DR, principally by imaging. A comprehensive review of computational strategies for DR was published in 2013. A two-stage technique, diabetic fundus image recuperation (DFIR), has been used to predict DR. The first stage includes an analysis of advanced retinal fundus images. The subsequent stage uses a help vector machine for the forecast. This resulted in an approach to determine a patient's need for referral based on evidence of DR-related injuries on retinal imaging.

7.5.2 Classification

For semi-supervised learning strategies, the objective is to create a model that results in one of a variety of potential classes using named and unlabeled datasets. The easiest class of semi-supervised learning methods trains one model using one learning calculation with many characteristics. For example, self-preparing first trains a model on the marked models. The model is then applied to all unlabeled information, whereby the models are positioned by the confidence in their predictions. The most certain predictions are then included in the named models. This procedure repeats until all unlabeled models are marked.

Another class prepares different classifiers by inspecting the preparation information several times and preparing a model for each example. For example, tripreparing first trains three models on the marked models using bagging for the outfit learning calculation. At that point, each model is refreshed iteratively, whereby the other two models make predictions on the unlabeled models. Only models with similar forecasts are used to with the first named guides to re-train the model. The cycle stops when no model changes. Finally, the unlabeled models are marked utilizing lion's share casting a ballot in which at any rate two models must concur with one another.

7.6   Summary

This chapter focused on predictive analytics with machine learning to analyze Big Data in order to predict future complications in diabetic patients. A dataset of 4.101 million people and 42,000 factors from case management, pharmacy records, medical records, and laboratory results between 2005 and 2009 was used to create prescient models (to calculate relapse) for various onsets of T2D. The most certain predictions were then included in the marked models. This procedure repeated until all unlabeled models were named. The following class prepared different classifiers by inspecting the preparation information several times and preparing a model for each example. At that point, each model was refreshed iteratively, whereby the other two models made predictions on the unlabeled models; only the models with similar forecasts were used with the first named guides to re-train the model. The cycle stopped when no model changes. Finally, the unlabeled models were named utilizing dominant part casting a ballot in which at any rate two models must concur with one another.

References

1. American Diabetes Association. (2009). Diagnosis and classification of diabetes mellitus. *Diabetes Care, 32*(Suppl. 1), S62–S67.
2. Bagherzadeh, F., Khiabani, A., Ramezankhani, F., Azizi, F., Hadaegh, E. W., & Steyerberg, D. (2016, March). A tutorial on variable selection for medical prediction models: Feature selection methods in data mining could improve the results. *Journal of Clinical Epidemiology, 71*, 76–85. https://doi.org/10.1016/j.jclinepi.2015.10.002.
3. Agrawal, R., & Srikant, R. (1994). Fast algorithms for mining association rules in large databases. In *Proceedings of the 20th international conference on very large databases* (pp. 478–499).
4. Cai, L., Wu, L., Li, D., Zhou, K., & Zou, F. (2015). Type 2 diabetes biomarkers of human gut microbiota selected via iterative sure independent screening method. *PLoS One, 10*(10), e0140827. https://doi.org/10.1371/journal.pone.0140827.
5. Haldorai, A. R., & Murugan, S. Social aware cognitive radio networks. In *Social network analytics for contemporary business organizations* (pp. 188–202). https://doi.org/10.4018/978-1-5225-5097-6.ch010
6. Fayyad, U., Piatetsky-Shapiro, G., & Smyth, P. (1996). From data mining to knowledge discovery in databases. *AI Magazine, 17*, 37–54.
7. Abawajy, J., Kelarev, A., Chowdhury, M., Stranieri, A., & Jelinek, H. F. (2013). Predicting cardiac autonomic neuropathy category for diabetic data with missing values. *Computers in Biology and Medicine, 43*(10), 1328–1333. https://doi.org/10.1016/j.compbiomed.2013.07.002.

8. Agrawal, R., Imielinski, T., & Swami, A. (1993). Mining association rules between sets of items in large databases. In *Proceedings of the ACM SIGMOD conference on management of data* (pp. 207–216).

9. Georga, E. I., Protopappas, V. C., Polyzos, D., & Fotiadis, D. I. (2015). Evaluation of short-term predictors of glucose concentration in type 1 diabetes combining feature ranking with regression models. *Medical & Biological Engineering & Computing, 53*(12), 1305–1318. https://doi.org/10.1007/s11517–015–1263-1.

10. Anandakumar, H., & Umamaheswari, K. (2017, September). An efficient optimized handover in cognitive radio networks using cooperative spectrum sensing. *Intelligent Automation & Soft Computing*, 1–8. https://doi.org/10.1080/10798587.2017.1364931.

11. Huang, G.-M., Huang, T.-Y., & Lee, J. (2015). An interpretable rule-based diagnostic classification of diabetic nephropathy among type 2 diabetes patients. *BMC Bioinformatics, 16*(S-1), S5.

12. Jelinek, H. F., Stranieri, A., Yatsko, A., & Venkatraman, S. (2016). Data analytics identify glycated haemoglobin co-markers for type 2 diabetes mellitus diagnosis. *Computers in Biology and Medicine, 75*, 90–97. https://doi.org/10.1016/j.compbiomed.2016.05.005.

13. Lagani, V., Chiarugi, F., Thomson, S., Fursse, J., Lakasing, E., Jones, R. W., et al. (2015, May–June). Development and validation of risk assessment models for diabetes-related complications based on the DCCT/EDIC data. *Journal of Diabetes and its Complications, 29*(4), 479–487. https://doi.org/10.1016/j.jdiacomp.2015.03.001.

14. Lagani, V., Chiarugi, F., Manousos, D., Verma, V., Fursse, J., Marias, K., et al. (2015, July). Realization of a service for the long-term risk assessment of diabetes-related complications. *Journal of Diabetes and its Complications, 29*(5), 691–698. https://doi.org/10.1016/j.jdiacomp.2015.03.011.

15. Lee, B. J., & Kim, J. Y. (2016). Identification of type 2 diabetes risk factors using phenotypes consisting of anthropometry and triglycerides based on machine learning. *IEEE Journal of Biomedical and Health Informatics, 20*(1), 39–46. https://doi.org/10.1109/JBHI.2015.2396520.

16. Leung, R. K., Wang, Y., Ma, R. C., Luk, A. O., Lam, V., Ng, M., et al. (2013). Using a multi-staged strategy based on machine learning and mathematical modeling to predict genotype–phenotype risk patterns in diabetic kidney disease: A prospective case–control cohort analysis. *BMC Nephrology, 14*, 162. https://doi.org/10.1186/1471-2369-14-162.

17. Anandakumar, H., & Umamaheswari, K. (2017, March). Supervised machine learning techniques in cognitive radio networks during cooperative spectrum handovers. *Cluster Computing, 20*(2), 1505–1515.

18. Marling, C. R., Struble, N. W., Bunescu, R. C., Shubrook, J. H., & Schwartz, F. L. (2013). A consensus perceived glycemic variability metric. *Journal of Diabetes Science and Technology, 7*(4), 871–879.

19. Marx, V. (2013, June 13). Biology: The big challenges of big data. *Nature, 498*(7453), 255–260. https://doi.org/10.1038/498255a.

20. Mitchell, T. (1997). *Machine learning* (p. 2). Singapore: McGraw Hill. ISBN:0-07-042807-7.

21. Haldorai, A. R., & Murugan, S. Social aware cognitive radio networks. In *Social network analytics for contemporary business organizations* (pp. 188–202). https://doi.org/10.4018/978-1-5225-5097-6.ch010.

22. Sacchi, L., Dagliati, A., Segagni, D., Leporati, P., Chiovato, L., & Bellazzi, R. (2015). Improving risk-stratification of diabetes complications using temporal data mining. In *Conference proceedings – IEEE engineering in medicine and biology society* (pp. 2131–2134). https://doi.org/10.1109/EMBC.2015.7318810.

23. Stranieri, A., Abawajy, J., Kelarev, A., Huda, S., Chowdhury, M., & Jelinek, H. F. (2013). An approach for Ewing test selection to support the clinical assessment of cardiac autonomic neuropathy. *Artificial Intelligence in Medicine, 58*(3), 185–193. https://doi.org/10.1016/j.artmed.2013.04.007.

24. Wang, K. J., Adrian, A. M., Chen, K. H., & Wang, K. M. (2015). An improved electromagnetism-like mechanism algorithm and its application to the prediction of diabetes mellitus. *Journal of Biomedical Informatics, 54*, 220–229. https://doi.org/10.1016/j.jbi.2015.02.001.

Chapter 8
On-the-Go Network Establishment of IoT Devices to Meet the Need of Processing Big Data Using Machine Learning Algorithms

S. Sountharrajan, E. Suganya, M. Karthiga, S. S. Nandhini, B. Vishnupriya, and B. Sathiskumar

8.1 Introduction

With the advancement of technologies, a lot of devices are coupled together, generating data, and that data is very much bigger and complex. This huge volume of data is termed as "big data." Conventional method of data processing cannot manage such huge voluminous data. A lot of business-related queries are answered using this massive collection of data sets. So, this big data has to be processed in some manner to solve our business problems. Recent developments in open-source platform, namely, Spark and Hadoop, paved a way to handle and store this big data in a cheaper and more efficient way. Machines apart from humans are solely responsible in the generation of this massive data which is increasing day by day. Also with the evolution of Internet of Things (IoT) and machine learning, the generation of massive data has skyrocketed. With the growth of big data increasing day by day, the usefulness of this data is just underpinning. Cloud computing with its scalability and elasticity expands the usefulness of big data and provides a path for the developers to build ad hoc clusters for small data subsets and to test it.

S. Sountharrajan (✉)
VIT Bhopal University, Bhopal, India
e-mail: s.sountharrajans@vitbhopal.ac.in

E. Suganya
Anna University, Chennai, India

M. Karthiga · S. S. Nandhini · B. Vishnupriya
Bannari Amman Institute of Technology, Sathyamangalam, India
e-mail: karthigam@bitsathy.ac.in; nandhiniss@bitsathy.ac.in; vishnupriya@bitsathy.ac.in

B. Sathiskumar
VIT University, Chennai, India
e-mail: sathiskumar.b@vit.ac.in

© Springer Nature Switzerland AG 2020   151
A. Haldorai et al. (eds.), *Business Intelligence for Enterprise Internet of Things*,
EAI/Springer Innovations in Communication and Computing,
https://doi.org/10.1007/978-3-030-44407-5_8

8.2   Business Use Cases of Big Data

The massive big data helps in addressing a wide range of business actions from customer feedback to good analytics. Some of the use cases are listed below.

8.2.1   New Product Establishment

Top companies use big data to understand the needs of their customers. By analyzing the past and latest services, key attributes are classified and predictive models are created by understanding the relationship between these previous and current attributes which in turn builds new products/services. Any plan to produce and launch new products is made after specific analytics of the data from focused groups, media, markets, and stores.

8.2.2   Predictions About Mechanical Failures

By analyzing structured and unstructured data hidden in organizations such as equipment manufactured year, model, entries of the log, errors, and engine and room temperature, mechanical failure of the machines can be predicted earlier before a problem breaks out. This prediction helps the company to reduce the amount spent for maintenance and to uplift their income.

8.2.3   Experience of the Customers

The experience of the customers can be known well than before with the availability of big data. With the help of social media, log files, web surfing, and other data, big data improves the customer experience knowledge and the delivery value. It helps in handling the issues positively and thereby diminishes the customer's stress.

8.2.4   Security Breaches

Security has become a big landscape nowadays. Complaints and security issues are growing constantly. Data patterns related to fraudulence are easily identified from big data and the valuable information is alienated to induce fast reporting.

8.3   Handing Huge Data

Each day 2.5 billion gigabytes of data are generated with an increase in complex patterns to interpret. To process all those, we need an intelligent system. For example, data from social media and sensory information collected by satellite in space need to be recorded for analysis. The following primitive systems can be utilized to break the enormous amount of data, which takes more time for human analysis [15].

Intelligent learning – is a journey, which indicates maturity level and capability of automation of the business process to improve efficiency and reduce operational cost. The system can do, think, learn, and adapt which in turn lead to artificial intelligence.

Artificial intelligence – is where a machine performs the task that requires human judgment, learning, or problem-solving skills using techniques like NLP, speech recognition, machine learning, and many more.

Machine learning – is a technique to implement artificial intelligence, which uses algorithms to understand patterns in a huge amount of data.

Deep learning – is a technique of learning complicated patterns in a huge amount of data using different types of neural networking concepts.

Cognitive learning – is a combination of all these techniques to automate a task that involves unstructured content decision-making or any complex challenge.

Among these buzzing techniques, the widely used techniques are machine learning and cognitive computing or learning.

8.4   Cognitive Learning (CL)

Efficient steering through a flood of unstructured information requires a new era of computing called cognitive computing. Before understanding cognitive learning, one should first know the meaning of cognition, which means context and reasoning form the basis of the cognition system, which mimics humans' reason and the process through which they analyze information and draw an inference. Cognitive learning plays a vital role in fetching data from the rational business aspects and implements the solution in an efficient way [1]. CL is used in the field of analyzing social media data and is the perfect assistant for treating a serious medical condition as it speeds up interaction and decision-making accordingly. CL can merge data from various sources of information upon balancing context and conflicting evidence to suggest the best possible answers. To process the above-said data, cognitive learning includes some learning technologies that use data mining, pattern recognition, and NLP to mimic the way the human brain works, and these were used to achieve the learning systems. To enact the above-said technologies, cognitive learning uses its adjacent computing technique called machine learning.

Machine learning algorithms are used to perform typical tasks and solve different types of problems, which require a vast amount of structured and unstructured data [2]. A refined way of identifying patterns and processing data leads the cognitive learning system to move to the next level of anticipating new problems and modeling possible solutions. To attain all these capabilities, it must have the following five key attributes/features:

8.5 Machine Learning

Machine learning is a data-driven approach with a class of self-learning algorithms. In traditional programming, the data is given as an input and the output is predicted. In the machine learning approach, the data is provided and output is predicted by the machines; in turn, machines learn from the data, find the hidden insights, and create a model. Examples of machine learning in the industry are found in Google Maps and Facebook – face recognition, virtual personal assistant, speech recognition, natural language processing, and so on. The objectives of machine learning include greater accuracy, greater coverage of problems, greater economy in obtaining solutions, and greater simplicity of knowledge representation. Machine learning, like human learning, has two aspects: a behavioral component and knowledge acquisition component [5].

The following are the features of the machine learning technique:

1. It adjusts the program action making use of the data to detect patterns in a data set.
2. It provides an efficient way of implementing complex programs.
3. It uses iterative algorithms to explicitly find the hidden perspectives.
4. It automates analytical model building.

A typical machine learning life cycle has six steps:

1. Collecting data – relevant data is collected from various sources.
2. Data wrangling – it is the process of cleaning and converting data into a specific format.
3. Analyzing data – it is the process of selecting and filtering the required data using machine learning algorithms.
4. Train and test algorithm – use the appropriate algorithm.
5. Prediction – a new set of input is given during testing and the model will classify the input according to the training data set.

8.6   Cognitive Computing and Machine Learning

Cognitive computing and machine learning can be used to support decision-making, deliver highly relevant information, and optimize the available attention to avoid missing key developments. Cognitive machine learning can perform operations analogous to learning and decision-making in humans [6]. Intelligent personal assistants can recognize voice commands and queries, respond with information, or take desired actions quickly, efficiently, and effectively.

Implementing cognitive computing can help achieve these desirable results:

- It helps people make better decisions, take action more quickly, and achieve more successful outcomes.
- It delivers relevant information and advice at the time of need.
- It reduces information overload and optimizes people's available attention span.
- It allows people to act more efficiently and effectively.
- It reduces errors, minimizes loss and damage, and improves health and safety.

Cognitive computing tools such as IBM Watson, artificial intelligence tools such as expert systems, and intelligent personal assistant tools such as Amazon Echo, Apple Siri, Google Assistant, and Microsoft Cortana can be used to extend the ability of humans to understand, decide, act, learn, and avoid problems. Using these approaches, one can enhance the capabilities of humans by augmenting their powers of observation, analysis, decision-making, processing, and responding to other people and to routine or challenging situations. Cognitive learning principles are ways to apply sentient thought to learning activities and in turn control learning behavior and motivate toward more profitable results.

The following are the three basic cognitive principles of learning strategies [3]:

1. Effective learning – rather than responding to stimuli, its focus is on what you know.
2. Emphasize structure – connect new information and sort the same.
3. Effective learning strategy – validate the information and act accordingly.

8.7   Cognitive-Machine Learning Algorithms

Machine learning deals with artificial intelligence algorithms to enable the computer to think and learn as efficiently as humans. The main goal of machine learning algorithm over cognitive computing is to design the process that helps in recognizing patterns and compose decisions based on input data. Machine learning is applied for training cognitive networks.

One of the best machine learning techniques is supervised learning which leads cognitive learning to work efficiently [4]. The feature of machine learning algorithm design depends on the appropriate selection of one or more basic principles, which are selected based on its learning tasks. The combination of these basic principles is suitable for learning tasks if only a short time is available for the solution and if an optimal solution is not necessary, which should be solved by the algorithm. Ordered version space search with the aid of score function and reduction of the number of concept versions solves a convoluted learning task with a large number of training examples that include the following:

8.7.1 Particle Swarm Optimization (PSO)

In 1995, James Kennedy and Russell Eberhart first described a nondeterministic optimization technique named particle swarm optimization (PSO), which is a swarm-intelligence-based approximate technique. In the area of cognitive computing, optimization techniques can be used to find the parameters/objectives that provide the maximum (or minimum) value of a target function for classification algorithms such as artificial neural networks and support vector machines. These classification algorithms often require the user to define certain coefficients, which are to be found by trial and error or exhaustive search [9].

The following are the perceptions of particle swarm optimization:

- The PSO algorithm retains multiple hidden solutions at one time.
- Objective functions determine the fitness value at each iteration.
- At the search space, each solution is represented by a particle in the fitness.
- The particles "fly" or "swarm" through the search space and determine the maximum value returned by the objective function.

8.7.2 Bayesian Cognitive Learning Method

1. Bayes theorem is a linear classification to analyze feature probabilities and assumes feature independence. It learns and predicts very fast and it does not require lots of storage. The prevailing applications of Bayes algorithms are real-time prediction, multiclass prediction, text classification, recommendation system, cognitive systems, and image processing [7]. Cognitive machine learning is sympathetic in the best hypothesis h from some space H, given observed training data. Bayesian learning is relevant for two reasons:

   - Explicit manipulation of probabilities
   - Perspective for understanding learning methodology

2. The following is a feature of Bayesian learning methods:

   - Each observed training sample that could incrementally decrease or increase the estimated probability is correct.

3. Final probability of a hypothesis is determined from the prior knowledge from the observed data.

    • Weighted probabilities with combined predictions of multiple hypotheses were used to classify a new instance.

4. Bayes theorem and cognitive learning can be used for designing straightforward learning algorithms, which is achieved through a brute-force MAP learning algorithm. A Bayesian algorithm is a way to distinguish the behavior of learning algorithms. The advantage of using the Bayesian approach is that it uses independence assumption, which might not hold true in certain databases that could contain many interactions between the predictor (independent) variables. It has been applied to a variety of data mining analyses in a large number of domains.

8.7.3 Hill Climbing-GA (HCA-GA)

5. Hill climbing is a heuristic approach used for numerical optimization problems in the field of cognitive learning. The features of hill climbing include (1) variants of test algorithms as it takes the feedback from the test procedures. It is an iterative calculation that begins with an arbitrary answer for an issue, and after those attempts, it improves the answer by making changes. The cognitive engine has three major functions: optimization, decision-making, and learning.
6. HCA is good for finding a local minimum but not for a global solution [6]. Hill climbing is coupled with the genetic algorithm to find a global solution in the cognitive system. GA is a computer science search technique used to discover approximate alternatives for issues of optimization and search.
7. In particular, it falls within the category of local search methods and is thus usually an unfinished search. HCA-GA is a method for hybrid optimization. In each genetic iterative, after the genetic operation, we get an ideal person from the entire population. Then the hill climbing algorithm optimizes the optimal individual again.

8.7.4 Associative Memory Algorithm

8. Associative memory is a framework of content-addressable memory that maps particular representations of inputs to particular depictions of output. Associative memories store content in such a manner that information can be recovered later by adding a tiny part of the material to the memory instead of giving the memory an address. Associative memories are used in database engine algorithms, anomaly detection systems, compression algorithms, cognitive learning, and face recognition systems as construction blocks. Hopfield's neural network is a classic illustration of associative memory [8].

It is a system that "associates" two patterns (X, Y) such that when one is encountered, the other can be recalled. Typically, $X Î \{-1, +1\}$ m, $Y Î \{-1, +1\}$ n and m and n are the length of vectors X and Y, respectively. The vector components can be thought of as pixels when the two patterns are considered as bitmap images. There are two associative memory classes: auto-associative and hetero-associative. An auto-associative memory is used to collect a pattern earlier stored that is most similar to the present pattern, i.e., X = Y. On the other hand, in a hetero-associative memory, the retrieved pattern is, in general, different from the input pattern not only in content but possibly also different in type and format, i.e., X ¹Y.

8.8 Big Data: A Challenge

Though big data offers plenty of use cases, there are lots of challenges in utilizing it. First and foremost the size of the data is big. In spite of the advancement in new technologies in storing and managing the data, the size of the data is increasing rapidly day by day. Yet organizations are trying to determine the best solution to store and retrieve their data in an efficient manner. Challenges not only occur in data storage but also in using the valuable information the data hold according to the needs. Data cleaning and data organization in a meaningful manner with respect to the needs of the clients are still a challenging task. Nearly 60–70% of the time is spent on the data preparation task itself. In spite of the challenges in storing and retrieving big data, the technologies used in handling the big data are varying rapidly. Apache Hadoop technology is utilized a few years back and then Spark technology is introduced. Presently both Apache Hadoop and Spark are used together.

8.9 Complexity of Big Data in Internet of Things (IoT)

One of the huge revolutions brought in the technological world is the Internet of Things (IoT) [10–12]. IoT helps to know about the things happening around us using sensors and to infer useful information from the sensors to utilize our world in a smart manner. Thus the life of the people has improved a lot in a smarter way using these real-world applications. Smart city is one of the use cases of IoT where a lot of applications operate together to make the city smart [13, 14, 16, 17]. The biggest challenge in IoT is its huge collection of data. Data collected from the applications cannot be utilized as it has to be processed to obtain useful information.

Three approaches are utilized in data processing of the big data collected from IoT applications:

- Cloud computing
- Edge processing
- Local computing

8.9.1 Local Computing

In this approach processing of data will be done in the place where data is composed. No data will be communicated to the servers located in remote places. Useful information if necessary to take advance decisions will only be communicated to remote places. This in turn enhances the whole efficiency and performance and also diminishes the added burden of transmission. The technology of local processing is used in smart sensors. Smart sensors are sensors with inbuilt computation and communication paradigm. A smart sensor collects data and then processes the collected data to make smart decisions and also store the data for future reference. It also provides two-way communication. It has become nowadays an integral tool in intelligent systems and an important perspective in advance IoT applications. One of the best examples is "smart wearables". These wearables obtain information from the environment, process and produce the user's required information, and also communicate the relevant information to the external platforms if needed.

The new implementation of intelligent systems in industries using IoT brought revolutionary changes thereby leading to the next phase of industrial revolution through "industrial IoT (IIoT)" [19]. Network virtualization plays a noteworthy role in providing flexible and manageable network connections [20]. Thus the intricacy of the infrastructure is reduced because the resources of the network are utilized as logical services rather than physical services. This facilitates the execution of setting up smart methodologies to calculate the network usability and data flows from an IoT application. But this resource utilization has to be properly carried out to enhance the effectiveness and efficiency. This is a challenging task to the network operators since active monitoring will produce an added overhead in the network. Yet, a promising methodology has to be implemented in an intelligent manner based on monitoring data partially and not as a whole.

8.9.2 Edge Processing

It is one of the emerging paradigms nowadays. It involves deploying storage and processing capability at the border or at the edge of the network. It is created between the data sensor area and the data cloud centers. Computation capability, storage ability, and resources for network are deployed. This capability of gaining computation capability near the data centers helps in attaining low latency, more bandwidth, and less jitter to services [21]. Some of the approaches that are implemented on account of edge processing are as follows:

1. Fog computing: It provides storage and computation capability by using fog nodes which consist of devices like routers, switches, and gateways of the network. These fog nodes with devices are considered as virtual nodes thereby contributing to network virtualization facility. This capability leads to the larger usage of fog processing in mobiles as well as in IoT devices.

2. Mobile edge computing (MEC): It involves deploying cloud computation in the base stations [22].
3. Cloudlets: It involves cloud computation facility without the facility of wide area networks (WANs). Local data acquisition network area contains the servers which are known as cloudlets.

8.9.3 Cloud Computing

It provides flexibility in yielding computation resources on pay basis. Currently, it is the most widely available disruptive technology. The data centers are deployed in large numbers and they are used by the users virtually with greater efficiency. This helps the usage of IoT use cases to operate in various environments in a lively manner without changing the infrastructure [40]. This facility increases the usage of IoT as a service [23]. It leads to efficient computation by integrating various IoT devices and other embedded systems thereby outsourcing advanced services and use cases based on collected data. The big data that is accumulated in the cloud is analyzed and useful information is gathered and outsourced to the stakeholders using data mining techniques. Complexity and constraint arising during operation are resolved and more favorable dynamic solutions are obtained.

8.9.4 Context-Aware Computing

Context-aware computing systems have a variety of functions like gathering context information from sensor devices and application interface and storing in a repository. Then context filtering of raw data is done to obtain useful information and model it into a useful context. This context information is utilized by the system to react and make appropriate decisions to the users via the interface. A recent advancement in context-aware computing is web-service-based context-aware systems [18]. A wide variety of middleware are also available in providing context-aware computations. Context-responsive applications that are aware about the different situations are designed as objects in RCSM [24]. The modeled objects gather data from different situations and accumulate it into object containers where analysis in respect to the situations is made. Context-aware applications are created using the infrastructure and programming framework Java Context Awareness Framework (JCAF) [25]. The issues related to big data analytics are also addressed by JCAF. User-preferred decision supporting system with a programming framework is provided by the middleware PACE [26]. Network-operated intelligent bots are represented in CAMPUS infrastructure for supporting context reasoning related to users and environment [27]. CoWSAMI [28] supports context-aware computing in pervasive environments. Context awareness related to disaster management is supported by web-service-related middleware ESCAPE [29]. It has both front end and back end.

Front-end components support context data sensing and data sharing for contextual information processing through web services. Back-end components support storage and processing of data from different front ends. Another web-service-related intensive working environment for context sensing and reasoning is provided by InContext [29]. Context-aware computing systems not only analyze and respond according to the behavior of the environment but also perk up the response of the software. It is a dynamic infrastructure with self-adaptable and configuring ability.

8.10   Issues in IoT Things and Big Data

Things in Internet of Things communicate with themselves and with surroundings for data aggregation and to respond unaccompanied to the real world without any human intrusion when taking important business decisions and social processes. It also creates services for interactions with the smart devices connected over the Internet and also changes their states when security breaches are accounted in maintaining the privacy of the users. The sensor devices in IoT operate anywhere and anytime. They gather different kinds of data for which it is meant for, like collecting humidity data for monitoring temperature, gathering noise level in case of environmental monitoring, obtaining biomedical data for patient monitoring system, etc. These data are big data and there are a lot of research issues related to the data collected. Sensor data are used by context-aware use cases to acclimatize to the changes and respond accordingly. Sensor data not only are complex, highly dynamic, and inaccurate and alter according to the time, but also these data have to be integrated in a repository and the information has to be analyzed specific to the domain.

To determine the complex relations between situations and data from sensor, appropriate machine learning techniques are employed. But the performance is highly dependable upon the complexity of the sensor data. Hidden Markov model (HMM) is utilized in most of the context-aware use cases where a Markov chain method (series of events) is used to model the system [30]. The system contains lot of finite hidden states and responses generated from these states. SVM (support vector machine) is a machine learning technique utilized for linear and non-linear data classification [31]. The protective nature of SVM against over-fitting helps in managing large feature spaces irrespective of the number of features.

With the wide knowledge of the environment and the needed context and how it is to be used, developers decide which features from data are needed to support their applications. But moving to the actual implementation from a design is a big task. Help from different services is needed by the developers to achieve this task. Webinos EU [32] is a funded project which provides a platform for a wide variety of wireless applications. Multi-platform, application-specific systems related to web technology are provided by Webinos for connected things. To do so, Webinos at any time makes use of the knowledge about the recent state of the IoT device and users to provide decisions based on the context. Thus Webinos is a good example in providing cross-platform environment for third-party context-aware services.

8.11   Opportunistic Data Distribution Maintenance for IoT

Access of data content from many devices has widely increased nowadays. The most widely accessed data by most of the people via mobile devices are short text messages, voice messages, and contacts. People access their SMS, mails, social network accounts, calendar data, photos, videos, music, games, and apps from anywhere and anytime using Wi-Fi. These accesses via wireless media produce an avalanche of data which has to be managed and filtered using big data platforms. In opportunistic networks, mobile devices are used for communication to achieve larger throughputs during failure of direct communication. These opportunistic networks are used for short-range communication when users are nearer to each other. The network routes are dynamically created. If a node wishes to transfer data, it first stores it and then passes the data across the network till it reaches the destination node that is in closer proximity to the sender node. From this node the data has to be relayed and this is one of the biggest challenges in opportunistic networks. A lot of solutions have been proposed in determining the suitable relay node such as choosing the highest centrality node [33], predicting the nodes that follow the same interaction pattern everyday [34], and deciding upon contact and inter-contact time of the node.

Data distribution is an imperative research area in opportunistic networks. In opportunistic networks, the topology is unstable due to its dynamic nature. New techniques based on design models in distributing data are proposed by different authors where data is distributed from the sources to the really interested receivers because of the unawareness of the sources and receivers previously. A lot of research fields have emerged based on delay-tolerant networks (DTNs) and opportunistic networks (ONs) for IoT use cases. These networks do not have a permanent infrastructure as like traditional networks whose nodes are assumed as permanent and topology is known beforehand. In traditional network, if a node likes to transmit a message, it embeds the message into the network path, and through the path, the message will reach its destination. In DTNs and ONs, the technique of message transfer is different. No two nodes are neighbors at a different time period. Nodes that are in close proximity can come in contact while there is data exchange. The buffer space is not available for the nodes in DTNs and ONs as they are mobile devices, so all the data cannot be exchanged at a time. It is relayed on store-carry-forward paradigm. These DTNs' and ONs' techniques are used mostly in IoT scenarios where the communication via traditional network approach is difficult and costly.

Nowadays online mobile advertisers generate revenues by targeting a users' wish and preference. For revealing appropriate ads based on users' preferences, precious information about the users is collected by these advertisers. The security and privacy of the users' data is not guaranteed. MobiAd is a solution provided for ensuring users' privacy by revealing the users' preferences to the advertisers yet preserving the users' privacy [35, 36]. An example of a context-aware mobile application solution is CAPIM (context-aware platform using integrated mobile services) [37].

It is a next-generation application integrated with a lot of services to gather context data. It uses the sensing ability of modern smartphones and other external sensors while dynamically loading the needed smart services. For smart city application, DTNs and ONs are the best methods. In smart city application, the sensors are meant for monitoring and integrating the significant infrastructure of the cities, for optimizing the usage of resources, and for planning future maintenance activities. The changes in the environment are guessed and reacted upon automatically by the sensors. The cumulative intelligence of cities is employed through smart city use case which connects the physical, economic, social, and business activities of the cities. The communication by smart city sensors is done through ONs and DTNs. The data collected from all these sensors are gathered and processed efficiently through big data platforms.

8.12   Data Distribution Techniques in Opportunistic Networks

Data distribution in an opportunistic network is carried out through four popular techniques. First, socio-aware overlay algorithm develops an overlay that consists of nodes with high centrality values and high visibility [38]. Before starting the communication, the technique assumes that an infrastructure exists and forms an overlay with visible nodes from each community sector. The nodes are interacted by starting gossiping after community detection. The gossiping distribution begins by sending the message to random node groups. The node centrality measurement unit is used to choose a hub in a network by the socio-aware overlay algorithm. Bluetooth and Wi-Fi devices are utilized for discovering the nodes. Subscribe approach or unsubscribe approach is followed by the socio-aware method. During communication, broker node information along with the centrality list and timestamp information is passed to the nodes for subscriptions or unsubscriptions. During change of a broker node due to loss of closeness, a new subscription list is exchanged. Updates about the new broker nodes are transferred to all the brokers. Community brokers help in propagating the information about subscriptions during gossiping. When a broker receives a publication, it forwards it to all the other brokers, and then all the brokers update their own publication list. When there are lots of members in its community, the information is flooded to all the members through proper channels.

The community detection methodology is provided by community-related algorithms such as socio-aware algorithm. Depending upon the knowledge of standpoint of each and every node in a network, the socio-aware algorithm classifies the nodes and builds a community structure. In this community structure, the first level of nodes belongs to the same community, and these nodes have a large number of contacts and a stable standpoint duration. The next level of nodes is familiar stranger nodes; these also have a larger number of contacts, but the standpoint duration of these nodes is short. The third level is completely stranger nodes which have a smaller number of contacts with shorter contact duration. Finally, the nodes with smaller contacts but with higher standpoint duration are friend nodes. Community detection of nodes is done in a decentralized manner due to the unreliable and unfixed structure of the opportunistic networks. Each node should determine its own local community. Two algorithms are there for community detection in distributed environments, namely, simple algorithm and k-clique algorithm. To identify the local community, the nodes interact with the nearby devices and run a detection algorithm. This methodology is used in the data exchange level between the nodes. Each node maintains the content detail about the nearby confront nodes, the duration of their contacts, and the identified local community. In a wireless environment, infrastructure is not possible, so a wireless podcasting method is used to facilitate the content distribution among the podcasting devices in that wireless range [39].

In resource-constrained networks, the content is made available to all the users in contact without high usage of resources by ContentPlace method [41]. This method utilizes the knowledge of relationships among the users to decide upon the area to position the user data. The method of design is based on two preferences: one is grouping the users according to their interests and the other one is based on their social relationships.

8.13   Research Path in Big Data Platforms for IoT

The actual research path leads to the evolution of Internet of Things which provides a huge amount of data by connecting all things in the environment thereby raising the demand requirements in a complex way. The increasing growth of data volume day by day requires a highly scalable platform for processing the data, for managing the traffic in the network, and for efficiently storing the data. For maintaining the connection between the things during poor connection of Wi-Fi/wireless links, good communication protocols are desired to enable the connection and to maintain the network traffic. For storing, processing, and manipulating the data in a proficient manner, new algorithms are needed. IoT helps in developing a lot of applications that connect the people and the things around the environment. They help the urban population in gathering information at their fingertips and also in mounting their activities to improve the competitiveness.

In addition to this, IoT also enables outdoor-computing facility and user purpose approaches to boost the lifestyle of the urban population like smart transportation, social services, anytime healthcare applications, agency administrations, and smart education. IoT is not a single technique; it is a combination of multiple technologies which tend to modify the society in the forthcoming years. The advancement of IoT depends on the growth of a huge number of research projects and journal publications. As estimated, the growth of IoT will accompany more than 50,000 billion things of different types. To accompany the growth of IoT, new protocols for communication and standard methodologies should be invented for communication between device and people. Web services can be used to embed the smart IoT devices and their applications. By increasing the consumption of resource time, response time, throughput time, and reliability, the quality of providing services could be enhanced. The availability of a lot of services and instant publishing/subscribing and notifying capability pave a path for easy management of the complex architecture of IoT.

The information obtained from IoT devices should be well processed so that the quality of the information is improved. The number of IoT devices is increasing, so proper management is required to provide a highly robust, autonomous, and intelligent operation. Solutions for self-organization, self-healing, self-protection, and self-optimization of IoT devices are needed. For distributed information storage and sharing, new computational services are available. For providing security and privacy to all IoT devices, new protocol mechanisms are essential. For context-aware IoT models, stronger security measures are essential. For saving the energy and for developing self-sustainable models, a new methodology is required.

A lot of research projects are going on to make the smart IoT objects gain their required energy from the environment itself, thereby facilitating smart energy management. In addition to this, new platforms and advanced techniques are essential for real-time storage and processing of big data to accomplish guaranteed data availability and provisioning in real time. One available technique for big data is cloud computing which simplifies the construction of big data infrastructure. It also facilitates massive big data storage and processing thereby satisfying the customer demands. New cloud computing capabilities are required for global collection of different kinds of big data and for distributed processing and secure communication among different platforms throughout the globe. When interoperability of IoT devices is considered, a huge variety of technologies and design procedures are involved. One approach is using standardized protocols for enabling inter-communication, and in addition to this, the self-configuring and self-managing abilities of smart devices are essential for inter-operability and inter-communication within the smart devices and with the surroundings. This methodology is highly preferable than centralized management of complex IoT infrastructure. The autonomous nature of IoT devices is preferred in their operational level. These automatic respondents of smart devices enable the building of complex infrastructure according to the changes in the environment. Special methodologies are needed for enabling low-power-consumption devices. Power management should be done starting at the operational level of devices to network routing level. This would enable us to develop complex expandable infrastructure at a low cost rate.

Due to the distributed nature of the surroundings, different issues may arise in the operations and decision-making of the collaborated devices. The major issue lies in device convergence and facilitating a quality solution. The other issue lies in aiming to provide a correct solution thereby avoiding duplication and malpractice. Altogether, things should be capable of preserving confidentiality, integrity, and data availability.

8.14   Summary

The evolution of Internet of Things leads to the growth of highly complex environments thereby raising demand requirements toward current and future research. The tremendous growth of data from IoT things requires highly reliable and scalable environments for supporting network traffic and storage and processing of data as needed by the consumers. For continuous connectivity of devices in wired and wireless links, standard communication protocols are essential to manage the high traffic in the network as well. In addition to this, new solutions are required for efficient storage, fetching, and searching of data in these complex environments. Such a solution would include forming dynamic networks of IoT devices based on the needs to process big data using any of the listed machine learning algorithms to provide useful information to the enterprise, which in turn helps in making the right decisions.

References

1. Ahmed, E., & Rehmani, M. H. (2017). Mobile edge computing: Opportunities, solutions, and challenges. *Future Generation Computer Systems, 70*, 59–63.
2. Anandakumar, H., & Umamaheswari, K. (2017, March). Supervised machine learning techniques in cognitive radio networks during cooperative spectrum handovers. *Cluster Computing, 20*(2), 1505–1515.
3. Anandakumar, H., & Umamaheswari, K. (2017, September). An efficient optimized handover in cognitive radio networks using cooperative spectrum sensing. *Intelligent Automation & Soft Computing*, 1–8. https://doi.org/10.1080/10798587.2017.1364931.
4. Arulmurugan, R., & Anandakumar, H. (2018). Early detection of lung cancer using wavelet feature descriptor and feed forward back propagation neural networks classifier. In *Lecture notes in computational vision and biomechanics* (pp. 103–110). https://doi.org/10.1007/978-3-319-71767-8_9.
5. Athanasopoulos, D., et al. (2008). CoWSAMI: Interface-aware context gathering in ambient intelligence environments. *Pervasive and Mobile Computing, 4*(3), 360–389.
6. Boldrini, C., Marco, C., & Andrea, P. (2008). ContentPlace: Social-aware data dissemination in opportunistic networks. In *Proceedings of the 11th international symposium on Modeling, analysis and simulation of wireless and mobile systems*. ACM, Vancouver, British Columbia, Canada.
7. Catherine, F., & Tamilarasi, J. S. (2018). Artificial intelligence – Deep learning-based cognitive ontology model. *International Journal of Innovative Technology and Exploring Engineering (ITITEE), 8*(2S2).
8. Celesti, A., et al. (2016). Exploring container virtualization in IoT clouds. In *2016 IEEE international conference on Smart Computing (SMARTCOMP)*. IEEE.
9. Ciobanu, R. I., & Ciprian, D. (2012). Predicting encounters in opportunistic networks. In *Proceedings of the 1st ACM workshop on high performance mobile opportunistic systems*. ACM.
10. Dobre, C., Flavius, M., & Valentin, C. (2011). CAPIM: A context-aware platform using integrated mobile services. In *2011 IEEE 7th international conference on intelligent computer communication and processing*. IEEE, Cluj-Napoca, Romania.

11. Doukas, C., et al. (2007). Patient fall detection using support vector machines. In *IFIP international conference on artificial intelligence applications and innovations*. Boston: Springer.
12. Giacobbe, M., et al. (2018). Evaluating information quality in delivering IoT-as-a-service. In *2018 IEEE international conference on Smart Computing (SMARTCOMP)*. IEEE, Taormina, Sicily, Italy.
13. Gil, D., et al. (2016). Internet of things: A review of surveys based on context aware intelligent services. *Sensors, 16*(7), 1069.
14. Haddadi, H., et al. (2011). Targeted advertising on the handset: Privacy and security challenges. In *Pervasive advertising* (pp. 119–137). London: Springer.
15. Haldorai, A., Ramu, A., & Murugan, S. Social aware cognitive radio networks. In *Social network analytics for contemporary business organizations* (pp. 188–202). https://doi.org/10.4018/978-1-5225-5097-6.ch010.
16. Henricksen, K., & Ricky, R. (2006). A survey of middleware for sensor networks: State-of-the-art and future directions. In *Proceedings of the international workshop on Middleware for sensor networks*. ACM, Melbourne, Australia.
17. Hui, P., Jon, C., & Eiko, Y. (2010). Bubble rap: Social-based forwarding in delay-tolerant networks. *IEEE Transactions on Mobile Computing, 10*(11), 1576–1589.
18. Kalyan, A., Srividya, G., & Sridhar, V. (2005). Hybrid context model based on multilevel situation theory and ontology for contact centers. In *Third IEEE international conference on pervasive computing and communications workshops*. IEEE, Kauai Island, HI, USA.
19. Kanda, T., et al. (2008). Who will be the customer?: A social robot that anticipates people's behavior from their trajectories. In *Proceedings of the 10th international conference on Ubiquitous computing*. ACM, Seoul, Korea.
20. Kim, H., Young-Jo, C., & Sang-Rok, O. (2005). CAMUS: A middleware supporting context-aware services for network-based robots. In *IEEE workshop on advanced robotics and its social impacts, 2005*. IEEE, Nagoya, Japan.
21. Kristina, M., & Jan, P. (2003). Basic principles of cognitive algorithms design. In *Proc. of the IEEE International Conference Computational Cybernetics, Siófok, Hungary* (pp. 245–247).
22. Lenders, V., Gunnar, K., & Martin, M. (2007). Wireless ad hoc podcasting. In *2007 4th annual IEEE communications society conference on sensor, mesh and Ad Hoc communications and networks*. IEEE.
23. Macke, J., et al. (2018). Smart city and quality of life: Citizens' perception in a Brazilian case study. *Journal of Cleaner Production, 182*, 717–726.
24. March, H. (2018). The Smart City and other ICT-led techno-imaginaries: Any room for dialogue with Degrowth? *Journal of Cleaner Production, 197*, 1694–1703.
25. Mayank, A., Joshua, C. P., & Thomas, L. G. (2019). Using Machine Learning to guide cognitive modeling: A case study in moral reasoning. *arXiv preprint arXiv*:1902.06744.
26. Mohammadi, M. (2018). *Enabling cognitive smart cities using big data and machine learning: Approaches and challenges*. IEEE.
27. Ng, I. C. L., & Susan, Y. L. (2017). The Internet-of-Things: Review and research directions. *International Journal of Research in Marketing, 34*(1), 3–21.
28. Saarikko, T., Ulrika, H. W., & Tomas, B. (2017). The Internet of Things: Are you ready for what's coming? *Business Horizons, 60*(5), 667–676.
29. Samir, M. (2017). *Cognitive computing architectures for machine (deep) learning at scale*. MDPI, Presented at the IS4SI 2017 Summit DIGITALISATION FOR A SUSTAINABLE SOCIETY, Gothenburg, Sweden.
30. Satyanarayanan, M. (2017). The emergence of edge computing. *Computer, 50*(1), 30–39.
31. Su, Y., et al. (2018). Dynamic virtual network reconfiguration method for hybrid multiple failures based on weighted relative entropy. *Entropy, 20*(9), 711.
32. Tomáš, K., Štěpán, B., & Johannes, F. (2018, April). A review of possible effects of cognitive biases on the interpretation of rule-based machine learning models. arXiv:1804.02969v1 [stat. ML].

33. Truong, H., & Schahram, D. (2009). A survey on context-aware web service systems. *International Journal of Web Information Systems, 5*(1), 5–31.
34. Van Kasteren, T., et al. (2008). Accurate activity recognition in a home setting. In *Proceedings of the 10th international conference on Ubiquitous computing*. ACM.
35. Vergori, P., et al. (2012). The webinos architecture: A developer's point of view. In *International conference on mobile computing, applications, and services*. Berlin/Heidelberg: Springer.
36. Williams, J. A., Weakley, A., & Cook, D. J. (2013). *Machine learning techniques for diagnostic differentiation of mild cognitive impairment and dementia.* AAAI Workshop.
37. Yau, S. S., & Fariaz, K. (2004). A context-sensitive middleware for dynamic integration of mobile devices with network infrastructures. *Journal of Parallel and Distributed Computing, 64*(2), 301–317.
38. Yingxu, W. (2017). *Novel machine learning algorithm for cognitive concept elicitation by cognitive robots.* IJCINI.
39. Yoneki, E., et al. (2007). A socio-aware overlay for publish/subscribe communication in delay tolerant networks. In *Proceedings of the 10th ACM symposium on modeling, analysis, and simulation of wireless and mobile systems*. ACM, Chania, Crete Island, Greece.
40. Zhihan, L., et al. (2018). Government affairs service platform for smart city. *Future Generation Computer Systems, 81*, 443–451.
41. Zhong, R. Y., et al. (2017). Intelligent manufacturing in the context of industry 4.0: A review. *Engineering, 3*(5), 616–630.

Chapter 9
Analysis of Virtual Machine Placement and Optimization Using Swarm Intelligence Algorithms

R. B. Madhumala and Harshvardhan Tiwari

9.1 Introduction

Virtualization is one of the powerful techniques that introduced a wide variety of applications to the real world. Cloud computing is considered as a computer paradigm for computing and delivering services over the Internet with many exciting features and provides three major service frameworks such as the software as a service (SaaS), the infrastructure as a service (IaaS), and the platform as a service (PaaS). Consumers are always concerned about the performance of the application that they are using and are very much interested in providing efficient resources. The predominant issues in effective cloud service provisioning are optimal resource utilization and energy conservation over the network. Due to their emerging growth, cloud providers also provide infrastructure as a service (IaaS) to their main operations. In cloud service models, the IaaS model provides virtualized hardware resources where a single physical server is partitioned into multiple logical servers; in turn, each logical server acts as a single physical server for problem execution.

The total energy used and the cost utilization in cloud services mainly depend on virtual machine scheduling. Efficient resource management through virtual machine placement is a great concern in data centers. Allocating resources for cloud data centers to formulate a virtual machine from a physical machine makes it a big problem. While allocating, there is a need to consider the basic parameters to optimize resource allocation. We are mainly concentrating on the energy consumption as well as the cost minimization of cloud data centers.

R. B. Madhumala (✉)
Department of Computer Science Engineering, Jain University, Bangalore, India

H. Tiwari
CIIRC, Jyothy Institute of Technology, Bangalore, India

© Springer Nature Switzerland AG 2020
A. Haldorai et al. (eds.), *Business Intelligence for Enterprise Internet of Things*,
EAI/Springer Innovations in Communication and Computing,
https://doi.org/10.1007/978-3-030-44407-5_9

The cloud service providers provide a group of computational memory and storage units in the form of virtual machines. Each physical bare-metal unit is divided into a number of VMs. The main concern of the cloud provider is to identify a proper physical machine to deploy the VMs meant at reducing the physical machine's running while maintaining service quality. This process is termed as virtual machine placement optimization.

The researchers continuously carried out the effective optimization of VMs. Few kinds of research used heuristic methods while others used metaheuristic methods with a flavor of artificial intelligence.

Optimization helps to completely utilize all the resources of a currently running physical machine (Fig. 9.1). Optimization helps in reducing power consumption thereby improving the green computing. In real-life situations, there will be many constraints for optimizing the cloud center. Maximizing the efficiency of the cloud center and minimizing the undesired factors is a challenging task to solve.

Optimization is a trial and error method. Many new algorithms will be proposed and their results are tested against the desired metrics and the successful ones are continuously modified to get better results. Broadly we can classify the optimization algorithms as conventional and non-conventional algorithms. Much of the work is already done in the conventional algorithms, but very few are proposed in non-conventional algorithms. The non-conventional algorithms are primarily the bio- or nature-inspired algorithms. These nature-inspired algorithms need intelligence and the AI tools are used for inducing the required intelligence into the algorithms. Few

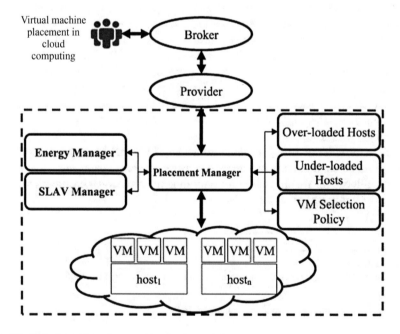

Fig. 9.1 Virtual machine placement in cloud computing

optimization algorithms are listed under heuristic methods as well as metaheuristic methods. In virtual machine selection, the overloaded host is detected and VMs are further selected to offload the host mainly to address the performance issue [1]. After selecting VMs, the host once again is checked if it still remains overloaded. In the next step, it is applicable to select another VM, which is purposed to migrate from its host [2]. This process is thus repeated unless the host is considered as unoverlooked. The aspect of virtualization is a vital concept of cloud computing which determines the manner in which a suitable host for the virtual machine is analyzed. As indicated above, the inspiration behind the placement of VM is inscribed in the traffic-aware, application-aware, topology-aware, and energy-aware prospects in Fig. 9.2.

Energy management—minimizing the cost at the level of hardware and server consolidation: green computing. Resource management—resources should be available based on the need and cannot be allocated statistically based on the peak workload elasticity of the cloud traffic engineering: to maintain the data center application efficiency and also for accurate planning of the network architecture.

Resources are limited to cloud data centers simply because cloud providers try to minimize over-utilized resources. This helps them to have a minimum number of servers as well as reduce the cost for both server hardware and software. Resources like CPU, memory, and hard disk need to be kept at minimum to avoid unnecessary costs. So some improvements are needed to reduce energy cost. This research has analyzed the recent researches concerning the provision of resources in the process of virtualizing computing ecological segment. It reduces response time and maximizes resource utilization but QoS factors are not considered [3]. It achieves schedule with a lower makespan, but its time complexity is more. It minimizes the total cost and high risk in resource management. Existing methods cannot operate on the issues of non-coordinate framework like resolution to power field.

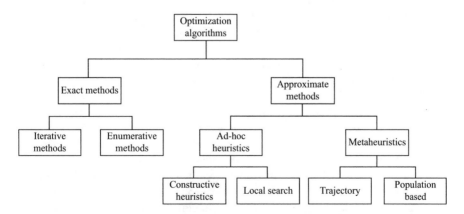

Fig. 9.2 Optimization algorithm classification

9.2 Literature Survey

This section provides insight into existing algorithms developed which can be used for optimal virtual selection of machines and the inclusivity of cloud computing, which is placed in the group of issues known as the NP-hard issues because of large resolution spaces. Cloud computing includes tasks for mapping on an unlimited computing resource, nature-inspired algorithm in ant colony optimization (ACO), particle swarm optimization (PSO), and gravitational search algorithm (GSA), which are vital for dealing with the NP-hard issues mentioned in this chapter. The condition-motivated algorithm in the area of privacy is meant to safeguard the data connected with cloud computing and task computing.

- Kumar et al. [4] presents a critical survey of nature clever algorithms that are used in artificial intelligence and automation in real-life domains. Nature-inspired algorithms are an emerging area of research on an algorithm based on physics and biology. This paper explains the idea of duplication in artificial frameworks. The condition-inspired computing and computation intelligence will assure maximum resolution to issues on a new venue of development and research.
- Nature with its massively adverse, robust, dynamic, and complex idea is a great source of information for solving problems in computer science. Biology-inspired computing has a wide field for research; in particular, there are great opportunities in exploring a new approach. Nature-inspired algorithm is a vital algorithm which increases the field of future-generation computing. Binitha et al. [5] presents a broad view of biology-inspired optimization algorithms which in turn increases the areas where these algorithms can be successfully applied.
- Cloud computing needs optimal resource utilization of cloud resources which in turn needs a certain novel scheme with enhanced dynamic resource allocation, collective control, resource management, and maximum distribution in networking and computing resources. The present trend is using virtualization in resource mobilization for data centers through machine migration techniques. Theja and Babu [6] discusses proposed approaches for the optimization of resources for cloud infrastructure. Virtual machines in association with physical machines can be an effective solution for resource optimization by considering certain increased predictive schemes, load balancing, and mapping which could be an increasing factor in virtualization for better performance.
- The VM placement in the cloud is enhanced based on objects such as VM time allotment, power consumption, SLA violation utility of resources, etc. [7]. This includes proposed algorithms such as the multiplied hybrid ACO-PSO algorithms that minimize the resource wastage and consumption of power to assure a more balanced server. It helps in the reduction of server costs.
- Usmani and Singh [8] deals with details regarding VM placement algorithm aiming at maximum utilization to reach optimal solution for minimization of power consumption. These algorithms aim at studying the workload variability and transforming the application demands, including the minimization of trade-off

between power consumption and better performance by using a hybrid technique for server energy efficiency. This is designed to be a two-staged procedure that is comprised of green computing and overload avoidance.

- VM migration is a source-intensive process to address VMs' progressive demand which is effective for the CPU cycles, communication bandwidth, and cache memory capability. The continuous movement becomes essential in managing the efficiency of data centers and in smoothing the application service. Choudhary et al. [9] deal with the problem faced in VM migration by having them shifted while there is a progressive operation making it possible for the VMs to be changed with zero downtimes. This shows the various forms of content, which include migration of the CPU states, storage contents, and memory contents. This analyzes the post-copy, pre-copy, and hybrid method of VM migration. The VM migration methods are sub-divided into two categories: models and frameworks.

- Cloud computing has attained remarkable growth in every field; provisioning, scaling, and maintenance of applications are achieved and it serves in a breeze. Rani and Bhardwaj [10] focuses on task scheduling using ant colony optimization genetic algorithm, PSO, and GSA. This survey deals with task scheduling in cloud computing based on current information and sources to build a good mapping relationship between tasks and resources. It compares the ant colony optimization with other techniques to prove the former is better in comparison.

- The main goal of [11] is to provide an understanding of the present algorithms and approaches that ensure an effective VM placement in the contest of cloud identification and computing that will be applied in future systems. The sophisticated VM placement optimization aims to reduce work, power, and cost and prevent congestion of data flow. The migration of VM requires a secure connection between the source and target servers. This aims to make a way for further work to address this problem for establishing and managing better communication.

- Cloud computing which is an important development in sharing and pooling of resources over Internet services is still in its infancy to achieve improvement; much research is required in various directions: one is proceduring the objective of scheduling to trace enough resources. This is placed in the segment of problems known as the NP-hard problems. There is no algorithm that provides optimized remedies within the polynomial timeframe to resolve this issue. Metaheuristic-based methods provide some remedies within the speculated timeframe. The metaheuristic techniques like ACO, GA, and PSO and two novel approaches like the League Championship Algorithm (LCA) and bat algorithm (getting inspiration from the echolocation behavior of bats, Yang introduced the bat algorithm) form these techniques. The comparative framework of these algorithms is centered on metaheuristic methods utilized for optimizing methods, nature of obligations, and ecology where these algorithms are applied. Son and Buyya [12] proposes priority-aware VM allocation (PAVA) which uses network topology information to allocate VM on the host which is nearest to the requester of the resource. The priority of the task is also considered as a parameter.

9.3 Optimization Techniques

9.3.1 Ant Colony Optimization Algorithm (ACO)

The ant colony optimization (ACO) algorithm is considered as one of the most recent algorithms that is competitive to other forms of algorithms [13]. The ACO algorithm is developed based on its combination with the ABC algorithm. The hybrid algorithm is formulated to link up the advantages of the global research ability of the ABC algorithm and the localized search ability of the ACO algorithm alongside the merits provided by the maximum and minimum algorithms. The algorithm initially utilizes the advanced max and min algorithm meant to schedule requests, whereas balancing is performed by ACO that is enhanced by the ABC algorithm. The vital aim of using multiple-objective ant colony system algorithm for the virtual machine placement problem is to obtain a collection of non-dominated resolution, the Pareto set, which minimizes the complete resource wastage and the consumption of energy [14]. The projected algorithm is analyzed with the instances recognized in literature. The remedies are performed and compared to the present multi-purpose genetic algorithm and the two-objective algorithms referred to as the bin-packing algorithm and the maximum and minimum ant system algorithm in Fig. 9.3.

Whenever the ant has to locate the shortened path between the colonies to the feeding sources, the moving ant produces some pheromone on the ground, which is responsible for forming paths by a trail of substances. Whereas the isolated ant migrates at a fundamentally randomized way, the ant that encounters the laid trail may potentially determine and decide based on a high probability to embrace it. As such, this reinforces a trail with the pheromone. This inclusive behavior that pops up has created an autocatalytic condition whereby more ants following the trail are attracted thus becoming a followed path. The procedure is grouped based on a posi-

Fig. 9.3 Ant colony behavior in the search space

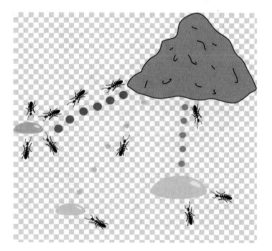

tive feedback loop, whereas the probability of the ant to select the path is developed through the number of ants which have chosen a single path.

9.3.2 Variants of Ant Colony Optimization (ACO)

Ant system: The initial ACO algorithm promised was the ant system (AS). This segment was applied to a number of small cases of travelling salesman issues in about 75 urban environments. On November 15, 1991, Dorigo submitted a manuscript for the ACO which was again revised on September 3, 1993, July 2, 1994, and December 28, 1994, and it was finally published on IEEE in 1996. It was an attempt by Dorigo to extend the travelling salesman (TSP) problem to the asymmetric travelling salesman problem (ATSP) [15].

Dorigo proposed the following algorithm:
1. Initialize the time counter, cycle counter, and trail intensity to zero.
2. Place m ants in n nodes.
3. Make the ants move to different towns with a probability function.
4. Each time a path is retraced, update the trail counter and prepare a tabu list with the values.
5. Steps 2 to 4 are repeated till the tabu list is full (all the towns are covered for each ant).
6. Steps 2 to 5 are repeated until the cycle counter reaches a preset value.
7. From the table, find the path with the highest trail intensity, and this is considered as the shortest path.

The complexity of the algorithms is determined by:

NC = number of cycles
n = number of towns
m = number of ants

Dorigo has even proposed two other extended versions of the algorithm, namely, ant-density and ant-quantity algorithms, but there was not enough research done as they didn't become more popular like the original ant system. The main problem of the ant colony algorithm is stagnation. When many ants travel in the problem space, a lot of pheromones will be put on that path, and since the path with a lot of pheromones is considered as the best path, there is a chance that the other available paths get ignored and this creates the stagnation in Fig. 9.4.

One issue which pops up based on the algorithms is stagnation. Whenever many ants travel in this issue space, they are placed down on pheromones. However, as many of them gravitate over the best resolution to locate it, it possibly creates starvation to the remaining datasets. Though few ants may take other paths, their pheromones will get evaporated eventually making that path as an unreliable path. This means that only the path where many ants travel will have pheromone replenish-

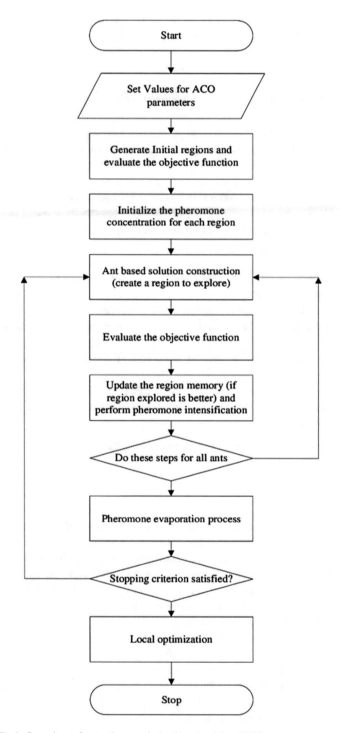

Fig. 9.4 Basic flow chart of ant colony optimization algorithm (ACO)

ment and all other paths will be ignored. When this happens, there is no possibility of further optimization, and to combat this, many modified and advanced versions of the ant system are engineered.

Elitist ant system: In the elitist AS, they have formulated the specialist ants alongside the normal ant system. In the elite ant system, the bonus pheromones are provided to the most relevant resolutions by the basis of multiplication of some pheromones provided by the specialist ants. At each moment, the ultimate path found is based on massive deposition of the pheromones that is initiated for a relevantly long duration of time. While the usual paths degrade as time goes, the elitist paths are rendered as a huge concentration of pheromones that stay and are a validated option for a long duration of time. As such, one elitist path is sufficient enough to diverge from the present issue space therefore preventing any possibility of stagnation.

Rank-based ant system: Ranked AS utilizes a form of a significant ant as seen in the case of the elitist ant system. But in this case, they spread the pheromones on various promising paths instead of providing one best path. Each path is ranked according to its length, whereby the best-ranked one gets more pheromones and the one that has been ranked the least obtains a few pheromones by the specialist ants. In the ranked AS, in case there are less specialists than the usual ants, the worst-ranked path shall not be assigned any pheromones.

Min-max ant system: In the minimum and maximum ant system, there exists no specialist ant. It is utilizing the normal ants. In this framework, there is a cap reflecting the max and min values of the pheromone, which can be placed on a certain path. Since the algorithm places the cap on a minimum pheromone amount, the path pheromone cannot be dropped as low to make the path so absolute. In the same case, the maximum amount of pheromone on paths is fixed as paths cannot get saturated in the process of overshadowing all the other paths. Whenever one or more paths are closer to the min or max levels, the pheromone levels are considered smooth. This therefore promotes the paths with low levels of pheromone while maintaining the standings over the paths. In this methodology, the pheromones are applicable on the most effective path while ignoring the others. All these additions make the basic ant system a powerful and competitive algorithm.

9.4   Particle Swarm Optimization Algorithm (PSO)

9.4.1   Introduction

In 1995, J. Kennedy and R. Eberhart based on their earlier investigations suggested that the group members can benefit not only from their unidirectional memories but also from the collective memory with a multidimensional group. Particle swarm optimization (PSO) method is based on bird flocking. It is a metaheuristic optimization algorithm. The major PSO algorithm operates based on a population or swarm of candidate resolutions known as particles. All these particles revolve around the issue search space as per the simple formula [16].

This is the concept behind the particle swarm optimization algorithm. We start with randomly grouped particles on the points of the forms $K = <k_1, k_2,...,k_n>$ and each is characterized with a randomized selected speed $V = <v_1, v_2,...,v_n>$. Every particle is composed of three elements such as inertia, memory, and group memories in Fig. 9.5.

The movements of the particles are guided by:

1. Their individually known positions in the searching spaces (*pbest*)
2. The complete swarm's best-known position (*gbest*)
3. The velocity V which is generally capped to some maximum value V_{max} to avoid searching out of the search space

Basic pseudo-code for PSO algorithm is provided below:

Step 1: Initialization

   For each particle i

   Initialize the particle's position with a uniform distribution and represent the lower and upper bounds of the search space.

(a) *Initialize pbest to its initial positions: pbest.*
(b) *Initialize gbest to the minimal value of the swarm: gbest.*
(c) *Initialize velocity: V.*

Step 2: Repeat until a termination criterion is met.

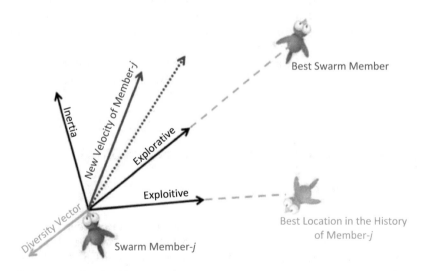

Fig. 9.5 Particles move around the problem search

For each particle i

Pick random numbers: r_1, r_2.
Update the particle's velocity. See formula (2).

(a) *Update the particle's position. See formula (3).*
(b) *Update the best-known position of particle i: pbest.*

Step 3: Output gbest(t) holds the best position found in a given problem search space.

The initial segment is termed as inertia that signifies the previous speed of particles. The following part is referred to as the cognitive element that is determined by every particle that signifies the other, including the third parameter that signifies the collaborative influence of the particles to locate the novel resolutions [17].

9.4.2 Basic Variants of PSO

Jau et al. [18] projected a modified QPSO that is applicable to the high-breakdown regression estimators and the minimally trimmed square methodology. Jamalipour et al. recommended the QPSO based on various mutation operators used to optimize the WWER-1000 core fuel processing. Tang et al. [19] projected a memetic algorithm and the memory technique. The memetic algorithm gives some basic experience to the particles in their local space, and after gaining experience in the local space, they are released to search the global space. Davoodi et al. [20] proposed a new approach, by combining improved QPSO and simplex algorithms. Li and Xiao [21] recommended the encoding technique that is centered on qubits described on the Bloch sphere. Yumin and Li [22] combined the artificial fish swarm to the QPSO and exploited adaptive constraints meant to be overlooked. Jia et al. [23] projected a developed QPSO centered on GA to evaluate the asynchronous optimization of the various sensor arrays and classifiers. Gholizadeh and Moghadas [24] focused and developed QPSO metaheuristic algorithm purposed to initiate a performance-based optimized designing process. The two number samples have been initiated to represent the most effective methodology. The two numerical examples have been included to represent the effectiveness of the proposed methodology. J. Kennedy proposed bare bones particle swarm optimization (BBPSO) in the year 2003. Modified the BBPSO algorithm by utilizing the crossover and mutation operator of the DE algorithms. In the year 2015 projected a binary BBPSO by formulating a reinforced memory technique to update all the local leader particles targeting to evade degradation of pending genes in particles to solve optimal feature subset and classification problems. Chaotic PSO (CPSO) was implemented by integrating the chaotic theory with the PSO. Many combinations were done with the CPSO algorithm like combining it with the K2 algorithm as done by Zhang et al. Pluhacek et al. [25] used various chaotic frameworks as a pseudorandom figure generator for the calculation of PSO algorithm. Juang et al. [26] projected the adaptive FPSO

(AFPSO) algorithm that uses fuzzy logic to enhance the model. PSO with time-varying acceleration coefficients (TVAC) was anticipated to enhance the performance of standard PSO by Cai et al.

9.5   Comparison of the Particle Swarm Optimization Algorithms

Author and year	Parameter used	Limitations	Advantages	Conclusion
Singh and Chana [26]	Execution time	It does not consider multiple levels of QoS requirements		QoS-aware resource scheduling in cloud environment
Dashti and Rahmani [27]	Makespan		It is more efficient for scheduling	The modified PSO algorithm
Lalwani et al. [28]	CPU-based and GPU-based strategies		Comparison between different PSO approaches	Study of variations of PSO algorithm
Xu and Yu [29]	Contraction-expansion coefficient, the wave function	It does not fully address the issue related to convergence rate and running time	Markov properties of SPSO are analyzed	The convergence of SPSO is studied using Martingale theory
Selvaraj et al. [30]	Turnaround time, waiting time, and CPU utilization		It is efficient when compared to SPSO, GA, and DPACO models	VM selection using swarm intelligence approach

9.6   Applications of PSO

Particle swarm is more than just a collection of particles wherein these particles have no power to solve any type of problem and progress occurs only when they interact with each other. The diagram represents the different applications of PSO in various domains as shown in Fig. 9.6.

PSO does not utilize the gradient of the issue but is centered on optimizing the necessary optimization issues that are defined by the classical optimization methodologies. PSO can be utilized in the process of optimizing the irregular issues. The US military is investigating swarm intelligence techniques for controlling unmanned vehicles. NASA is also investigating the use of PSO technology for planetary map-

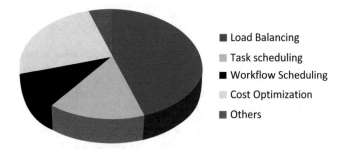

Fig. 9.6 Applications of PSO in cloud computing

ping. PSO is also applied in computer graphics and visualization. It is also used in designing and optimizing engines as well as electrical motors.

9.7 Summary

This chapter surveys different nature-inspired algorithms and their applications in different domains. We have also surveyed different algorithms used in optimal placement and selection of virtual machines in the cloud-centered ecological systems. All the algorithms discussed have considered various parameters like memory optimization, bandwidth optimization, computation time and cost, etc. As noted each algorithm has its limitations, and we need to develop a hybrid solution for optimizing the placement of virtual machines in the cloud.

References

1. Abdelsamea, A., et al.(2017, November). Virtual machine consolidation enhancement using hybrid regression algorithms. *Egyptian Informatics Journal, ,18*(3), 161–170.
2. Guo, Z., et al. (2017). Heuristic algorithms for energy and performance dynamic optimization in cloud computing. *Computing and Informatics, 36*, 1335–1360. https://doi.org/10.4149/cai201761335.
3. Dhanda, M., & Sangwan, O. P (2019). Quality of Service in Dynamic Resource Provisioning: Literature Review. In: Minz S., Karmakar S., Kharb L. (eds) Information, Communication and Computing Technology. ICICCT 2018. Communications in Computer and Information Science, vol 835. Springer, Singapore.
4. Kumar, M. S., Sindhuja, P., & Ramamoorthy, P. (2018). A brief survey on nature inspired algorithms: Clever algorithms for optimization. *Asian Journal of Computer Science and Technology, 7*, 27–32.
5. Binitha, S., & Sathya, S. S. (2012). A survey of bio inspired optimization algorithms. *International Journal of Soft Computing and Engineering, 2*(2), 137–151.
6. Theja, P. R., & Babu, S. K. (2014). Resource optimization for dynamic cloud computing environment: A survey. *International Journal of Applied Engineering Research, 9*(24), 26029–26042.

7. Suseela, B. B. J., & Jeyakrishnan, V. (2014). A multi-objective hybrid ACO-PSO optimization algorithm for virtual machine placement in cloud computing. *International Journal of Research in Engineering and Technology, 3*(4), 474–476.

8. Usmani, Z. & S. Singh (2016). "A survey of virtual machine placement techniques in a cloud data center." *Procedia Computer Science 78*: 491–498.

9. Choudhary, A., Govil, M. C., Singh, G., Awasthi, L. K., Pilli, E. S., & Kapil, D. (2017). A critical survey of live virtual machine migration techniques. *Journal of Cloud Computing, 6*(1), 23.

10. Rani, P. and A. K. Bhardwaj (2017). "A Review: Metaheuristic Technique in Cloud Computing." International Research Journal of Engineering and Technology (IRJET) Volume 4.

11. Stephanie, C., Paraiso, F., Merle, P. (2017). A Study of Virtual Machine Placement Optimization in Data Centers. 7th International Conference on Cloud Computing and Services Science.

12. Son, J., & Buyya, R. (2019). Priority-aware VM allocation and network bandwidth provisioning in Software-Defined Networking (SDN)-enabled clouds. *IEEE Transactions on Sustainable Computing, 4*(1), 17–28.

13. Blum, C. (2005). Ant colony optimization: Introduction and recent trends. *Physics of Life Reviews, 2*(4), 353–373.

14. Gao, Y., Guan, H., Qi, Z., Hou, Y., &Liu, L. (2013, December). A multi-objective ant colony system algorithm for virtual machine placement in cloud computing. *Journal of Computer and System Sciences,79*(8), 1230–1242.

15. Dorigo, M., & Gambardella, L. M. (1997). Ant colony system: A cooperative learning approach to the traveling salesman problem. *IEEE Transactions on Evolutionary Computation, 1*(1), 53–66.

16. Bin, W., Qinke, P., Jing, Z., & Xiao, C. (2012, June 1). A binary particle swarm optimization algorithm inspired by multi-level organizational learning behavior. *European Journal of Operational Research, 219*(2), 224–233.

17. Zhang, S., et al. (2017). Numerical study on attenuation of bubble pulse through tuning the air-gun array with the particle swarm optimization method. *Applied Ocean Research, 66*, 13–22.

18. Jau, Y.-M., Su, K.-L., Wu, C.-J., & Jeng, J.-T. (2013). Modified quantum-behaved particle swarm optimization for parameters estimation of generalized nonlinear multi-regressions model based on Choquet integral with outliers. *Applied Mathematics and Computation, 221*, 282–295. https://doi.org/10.1016/j.amc.2013.06.050.

19. Tang, D., Cai, Y., Zhao, J., & Xue, Y. (2014). A quantum-behaved particle swarm optimization with memetic algorithm and memory for continuous non-linear large scale problems. *Information Sciences, 289*, 162–189. https://doi.org/10.1016/j.ins.2014.08.030.

20. Davoodi, E., TarafdarHagh, M., & GhassemZadeh, S. (2014). A hybrid Improved Quantum-behaved Particle Swarm Optimization–Simplex method (IQPSOS) to solve power system load flow problems. *Applied Soft Computing, 21*, 171–179. https://doi.org/10.1016/j.asoc.2014.03.004.

21. Li, P. (2008). Quantum genetic algorithm based on Bloch coordinates of qubits and its application. *Control Theory & Applications, 25*(6), 985–989.

22. Yumin, D., & Li, Z. (2014). Quantum behaved particle swarm optimization algorithm based on artificial fish swarm. *Mathematical Problems in Engineering, 2014*, 1–10. https://doi.org/10.1155/2014/592682.

23. Jia, P., Tian, F., Fan, S., He, Q., Feng, J., & Yang, S. (2014). A novel sensor array and classifier optimization method of the electronic nose based on enhanced quantum-behaved particle swarm optimization. *Sensor Review, 34*(3), 304–311. https://doi.org/10.1108/SR-02-2013-630.

24. Gholizadeh, S., & Moghadas, R. (2014). Performance-based optimum design of steel frames by an improved quantum particle swarm optimization. *Advances in Structural Engineering, 17*, 143–156. https://doi.org/10.1260/1369-4332.17.2.143.

25. Pluhacek, M., Senkerik, R., & Zelinka, I. (2014). Particle swarm optimization algorithm is driven by multichaotic number generator. *Soft Computing, 18*, 631–639. https://doi.org/10.1007/s00500-014-1222-z.

26. Singh, S. and I. Chana (2015). "QRSF: QoS-aware resource scheduling framework in cloud computing." *The Journal of Supercomputing 71*(1), 241–292.
27. Dashti, S. E., & Rahmani, A. M. (2016). Dynamic VMs placement for energy efficiency by PSO in cloud computing. *Journal of Experimental & Theoretical Artificial Intelligence, 28*(1–2), 97–112.
28. Lalwani, S., H. Sharma, S. C. Satapathy, K. Deep and J. C. Bansal (2019). "A Survey on Parallel Particle Swarm Optimization Algorithms." *Arabian Journal for Science and Engineering 44*(4), 2899–2923.
29. Xu, G. and G. Yu (2018). "Reprint of: On convergence analysis of particle swarm optimization algorithm." *Journal of Computational and Applied Mathematics 340*, 709–717.
30. Selvaraj, A., R. Patan, A. H. Gandomi, G. G. Deverajan and M. Pushparaj (2019). "Optimal virtual machine selection for anomaly detection using a swarm intelligence approach." *Applied Soft Computing 84*, 105686.

Chapter 10
Performance Evaluation of Different Neural Network Classifiers for Sanskrit Character Recognition

R. Dinesh Kumar, C. Sridhathan, and M. Senthil Kumar

10.1 Introduction

Classifying the input characters according to the predefined character data set is a challenging task. Many researchers using the image processing techniques have contributed their work on English character recognition. There is an increasing interest for computer applications which ease the human jobs and solve complex problems. Optical character recognition is used for classifying the alphanumeric characters which are in the digital image format. Classical methods for handwritten character recognition have drawbacks. First, same characters will differ in style, shape, and size from person to person. Second, the characters will vary from time to time for the same person. Third, visual characters are affected with noise near the edges while reading through OCR techniques. Optical character recognition system has a variety of commercial and practical applications such as in banks, postal service, license plate recognition system, smart card processing system, and so on [1–3].

The third most commonly utilized language after English and Chinese is Hindi. There are around 500 billion people all over the world who write and speak in Hindi. Hindi is the fundamental script of many Indian languages that originated from Sanskrit. Sanskrit being an ancient language is no longer spoken. It is a very expressive language, which has been enriched and influenced by Farsi, Turkish, Dravidian, Portuguese, English, and Arabic.

Printed Sanskrit characters are effortlessly identified by computer machines. But, handwritten Sanskrit characters are not being identified accurately and

R. D. Kumar (✉)
Siddhartha Institute of Technology & Science, Hyderabad, Telangana, India

C. Sridhathan · M. S. Kumar
Nalla Malla Reddy Engineering College, Hyderabad, Telangana, India

© Springer Nature Switzerland AG 2020   185
A. Haldorai et al. (eds.), *Business Intelligence for Enterprise Internet of Things*,
EAI/Springer Innovations in Communication and Computing,
https://doi.org/10.1007/978-3-030-44407-5_10

efficiently by the computer machine. Numerous researchers have proposed different kinds of methods and algorithms for recognizing the Sanskrit characters. Several software are used for optical Sanskrit character recognition. However, for recognizing the handwritten Sanskrit characters, several procedures are to be adopted; no single machine or single process can perform Sanskrit character recognition. Artificial neural network (ANN) can be used for character recognition because of its versatility and simplicity of working design.

10.2 Literature Review

The functioning of the human brain has inspired many scientists to come up with computing systems termed as neural networks. Some researchers applied optical character recognition using feed-forward neural networks, backpropagation algorithm model, and other techniques for pattern recognition [1, 3, 4].

Recognition of handwritten characters is the most mesmerizing research area in the field of image processing and pattern recognition. In recent years, character recognition has been a popular research area for many researchers. This is because of its potential application in several fields like number plate recognition, sorting letter based on postal codes, etc. The first research on Devanagari characters was published in 1977 [5]. At present researchers are working on handwritten character recognition of many languages such as English, Chinese, Hindi, Tamil, Bangla, and Telugu. Few researchers are focusing on neural network and artificial intelligence technique for reduced processing time with higher recognition accuracy [1–3, 6–9]. Character recognition is mainly classified into online and offline which are further classified into clustering template matching, etc. as shown in Fig. 10.1 [8]. Existing

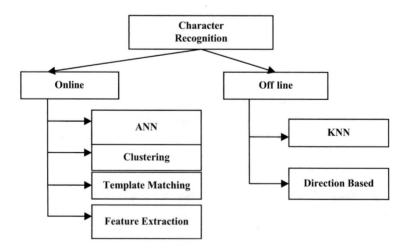

Fig. 10.1 Classification of character recognition system

methods for handwriting identifications are template matching, statistical and structural techniques, fuzzy logic, support vector machines, and neural networks (NNs).

Recognition of handwritten character involves (1) image acquisition, (2) image pre-processing, (3) image segmentation, (4) feature extraction, and (5) classification. Classification stage is the decision-making part of image processing systems. Feature extraction and classification play a significant role in handwritten character recognition [1, 2, 6].

10.3   Character Recognition Process

Character recognition process for offline handwritten patterns consists of the following stages as shown in Fig. 10.2.

10.3.1   Image Acquisition

The handwritten Sanskrit characters are scanned as an input as shown in Fig. 10.3.

10.3.2   Pre-processing

Scanned handwritten Sanskrit characters are transferred as an image. Pre-processing involves binarization, denoising, thinning, skew detection, and correction. In binarization, color images are converted into the gray-scale image with the help of the balanced histogram thresholding technique. Based on the histogram

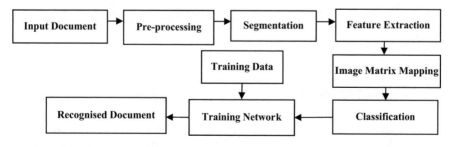

Fig. 10.2   Stages of Sanskrit handwritten character recognition system

Fig. 10.3   Scanned Sanskrit handwritten character

Fig. 10.4 Pre-processed Sanskrit handwritten character

Fig. 10.5 Character segmentation

weights, this method divides the images into two: foreground and background. Denoising is performed using non-local median filters for eliminating and suppressing the noise level. Zhang-Suen thinning algorithm has been utilized to produce another binary image with a thickness of one pixel for better edge detection. This minimizes the noises without losing the quality of the image. The final step in pre-processing is skew detection where the orientation of the character is adjusted with respect to the true horizontal axis. The angle of orientation of the character is rotated such that it is ±15 degrees. Pre-processed Sanskrit handwritten character is shown in Fig. 10.4.

10.3.3 Segmentation

In segmentation, the required region of the image is separated. Here, improved multi-scale segmentation approach is used to divide the characters into the horizontal and vertical projections with the help of the fragmentation process. After the fragmentation process, the texts are split into the paragraphs and then lines, words, and characters with the help of the histogram approach. Line, word, and character segmentation is carried out and segmented character images are shown in Fig. 10.5.

10.3.4   Feature Extraction

The statistical and structural procedure can be used for extracting features such as height, width, horizontal lines, vertical lines, circles, slope lines, and arcs. Here the height and width of the characters are estimated with the glyph and related boundary values. These detected line values are decomposed by applying the wavelet function described in Eq. 10.1.

$$F(a,b) = \int_{\infty}^{-\infty} f(x)\psi *(a,b)(x)dx \qquad (10.1)$$

where $*$ is the complex conjugated and ψ is the Daubechies wavelet function.

In addition mean, variance, standard deviation, entropy, and energy are also calculated and their corresponding formulae are given in Eqs. 10.2, 10.3, and 10.4.

$$Mean = \sum_{2(n-1)}^{i=0} i.p_{x+y}(i) \qquad (10.2)$$

$$Variance = \sum_{n-1}^{i=0} \sum_{n-1}^{j=0} (i-\mu)^2 .p(i,j) \qquad (10.3)$$

$$Entropy = \sum_{n-1}^{i,j=0} -\ln(p_{ij})p_{ij} \qquad (10.4)$$

10.4   Classification

Feature extracted characters are used in the classification stage for decision-making. The classification methods can be statistical methods, artificial neural networks (ANNs), kernel methods, and multiple classifier combination. In this work RCS with BPNN, BPNN with RBF, and multilayer perceptron model are used for classification.

10.4.1   RCS with BPNN

Random candidate selection (RCS) with backpropagation neural network (BPNN) is one of the efficient supervised learning models which maps each input to the particular output. The main goal of the RCS with BPNN is to train the network in an efficient manner that can learn the appropriate internal representations, which allows the arbitrary mapping of the input to the output with minimum error rate.

Steps involved in the proposed algorithm are as follows:

Step 1: Initialize all weights.

Step 2: Input vector consists of Fourier descriptors and border transition values which are used as classifiers.

Step 3: By using non-linear sigmoid function, the weights are adjusted to obtain the outputs yj.

Step 4: The weights are adjusted such that all the training inputs are used for weight stabilization.

The structure of the network and the training algorithm used in RCS improve the recognition of Sanskrit character and reduce the error rate based on the learning rate.

10.4.2 BPNN with RBF Network

The backpropagation neural network with radial basis function is one of the better classification algorithms with training process to get the desired output. RBFN consists of input layer, hidden layer, and output layer. Each layer produces the linear output even for non-linear inputs. In radial basis function neural network, extracted features like character height, weight, direction, mean, standard deviation, and variance are fed as inputs and transferred to the hidden layer that uses the radial basis (radbas) as the activation function. K-mean clustering algorithm is used to obtain the center value as given in Eq. 10.5.

$$argmin = \sum_{k}^{i=1} \sum_{X \in S_i} \|X - \mu_i\|^2 \tag{10.5}$$

In the recognition phase, the characters are identified based on the shape and the distance between the centers. Finally, the features of the handwritten character are stored as a template in the database and compared with the new character as given in Eq. 10.6.

$$y_k(x^p) = \sum_{M}^{j=0} W_{kj} \varphi_j(x^p) \tag{10.6}$$

10.4.3 Multilayer Perceptron Neural Network

Multilayer perceptron neural network's working principle is relatively based on the human brain. Normally a human brain stores the information as a pattern and gains the knowledge to solve the complex problems by experience. The multilayer perceptron neural network recognizes the patterns by supervised training algorithm with feed-forward from input to output layers. Activation function is computed as given in Eq. 10.7.

$$x_i^{m+1}(n) = f\left[\sum_M^{j=1...N^m} W_{ij}^m x_j^m(n)\right] \tag{10.7}$$

Based on the error rate, the weightage factor is modified as given in Eq. 10.8.

$$new.w_{ji}^m = w_{ij}^{m-1} + \gamma \sum_T^{t=1} \delta_i^m(n) x_j^{m-1}(n) \tag{10.8}$$

10.5   Performance Analysis

The performance of RCS with BPNN, BPNN with RBF, and multilayer perceptron neural network models is evaluated based on the following metrics:

10.5.1   Accuracy

Accuracy is a statistical measure which is used to analyze how well the classifier recognizes the Sanskrit character with optimized way in Eq. 10.9.

$$Accuracy = \frac{\text{No of True Positive} + \text{No of True Negative})}{\text{Total No of Samples}} \tag{10.9}$$

Accuracy can also be expressed in terms of sensitivity and specificity as given in Eq. 10.10.

$$Accuracy = (\text{sensitivity})(\text{prevalence}) + (\text{specificity})(1 - \text{prevalence}) \tag{10.10}$$

10.5.2   Sensitivity

Sensitivity is a measure of how the proposed system correctly classifies the handwritten characters with efficient manner. The sensitivity is measured using Eq. 10.11.

$$Sensitivity = \frac{\text{True Positive}}{\text{True Positive} + \text{False Negative}} \tag{10.11}$$

10.5.3 Specificity

Specificity measures how the proposed system correctly identifies the negative classifiers during the character recognition process. It is expressed as shown in Eq. 10.12.

$$\text{Specificity} = \frac{\text{True Negative}}{\text{True Negative} + \text{False Positive}} \quad (10.12)$$

The minimal error rate and precision recognition are used so that the efficiency of the Sanskrit character recognition system is increased.

10.6 Results

The experimental results of the random candidate selection with backpropagation neural networks, backpropagation neural networks with radial basis function networks, and multilayer perceptron neural network models are evaluated in terms of error rate, sensitivity, specificity, and accuracy measures. The results of the proposed classifiers are compared with the traditional classifiers such as SVM, SOM, etc., and the comparison results are shown below in Tables 10.1, 10.2, and 10.3.

Table 10.1 Performance evaluation of proposed RCS with BPNN and existing methods

Different classifiers	Sensitivity	Specificity	Accuracy
SVM	85	87	93
SOM	86.5	87.43	94.23
RBFN	88.23	91.23	94.34
Fuzzy NN	89.46	94.23	95.35
Proposed RCS with BPNN	92.13	95.63	96.45

Table 10.2 Performance evaluation of BPNN with RBF and existing methods

Different classifiers	Sensitivity	Specificity	Accuracy
SVM	85	86	92
SOM	83.5	86.43	93.23
RBFN	87.23	90.23	94.34
Fuzzy NN	88.46	93.23	95.35
BPNN with RBF	93.13	96.63	97.45

Table 10.4 Efficiency comparison for proposed character recognition methods

Metrics	RCS with BPNN	BPNN with RBF	MLP networks
Sensitivity	92.13	93.15	94.13
Specificity	95.63	96.63	96.73
Accuracy	96.45	97.45	98.46

Table 10.3 Performance evaluation of proposed MLP and existing methods

Different classifiers	Sensitivity	Specificity	Accuracy
SVM	86	87	97
SOM	87.5	86.42	96.89
RBFN	89.23	95.23	97.54
Fuzzy NN	91.46	91.43	97.77
Proposed MLP	94.13	96.73	98.46

The experimental metrics results of the MLP network models were compared with other models such as RCS with BPNN and BPNN with RBF; it had higher overall efficiency. The comparison results are shown in Table 10.4 below. It can be seen from Table 10.4 that the sensitivity, specificity, and accuracy levels of MLP level are higher. The methods used for recognizing the handwritten Sanskrit characters produce minimum error rate and high accuracy when compared to other existing methods.

10.7 Summary

Handwritten Sanskrit character recognition was done using the following methods: RCS with BPNN, BPNN with RBF, and MLP network. The MLP network gave an effective and efficient result in both feature extraction and recognition. The accuracy level of MLP level was 98% and execution time required was very less. This work can be extended in face recognition also.

References

1. Abbas, M. (2016). *Offline handwriting recognition using neural networks*. Thesis.
2. Aicha, E., Mohamed, K. K., & Hacene, B. (2015). Ontologies and bigram-based approach for isolated non-word errors correction in OCR system. *International Journal of Electrical and Computer Engineering (IJECE), 5*(6), 1458–1467.
3. Chirag, I. P., Ripal, P., & Palak, P. (2011). Handwritten character recognition using neural network. *International Journal of Scientific & Engineering Research, 2*(5), 1–6.
4. Khushbu, S. M. (2013). Image pre-processing on character recognition using neural networks. *International Journal of Computer Applications, 82*(13), 11–15.
5. Dimple, B., Gulshan, G., & Maitreyee, D. (2014). Design of an effective preprocessing approach for offline handwritten images. *International Journal of Computer Applications, 98*(1), 17–23.
6. Shoba, R., Sanjay, K. V., & Anitta, J. (2016). A zone based approach for classification and recognition of telugu handwritten characters. *International Journal of Electrical and Computer Engineering, 6*(4), 1647–1653.

7. Shobha, R., Neethu, O. P., & Nila, P. (2015). Automatic vehicle tracking system based on fixed thresholding and histogram based edge processing. *International Journal of Electrical and Computer Engineering, 5*(4), 869–878.
8. Shruti, S. K., Hoshank, J. M., Sakshi, S. G., Sarang, S. S., & Chitre, D. K. (2017). Handwriting recognition using neural network. *International Journal of Engineering Development and Research, 5*(4), 1179–1181.
9. Sudarshan, S., & Seema, B. (2016). Handwritten character and word recognition using their geometrical features through neural network. *International Journal of Application or Innovation in Engineering & Management, 5*(7), 77–85.

Chapter 11
GA with Repeated Crossover
for Rectifying Optimization Problems

Mayank Jha and Sunita Singhal

11.1 Introduction

Optimization is at the core of many processes that solve problems in the real world. Nonetheless, identifying the most optimal resolution for these issues is normally tiresome, mostly in the availability of non-linear, high-dimensionality, and multiple modalities. The evolving algorithm has indicated a more considerable success recorded over the past few years to deal with the complex issues of optimization. Whereas the EA family includes a number of various algorithms, the most prominent and commonly used in practice is the genetic algorithm (GA). There has been an increase in studies relating to the actual-parameter genetic algorithm (GA) in the past decades irrespective of the presence of one similar actual-parameter evolutionary algorithm, such as evolutionary technique and the differential evolution. In contrast, in theory, the actual-parameter GA studies indicated the same theoretical behavior in fitness landscapes with appropriate parameter tuning, as shown in the previous study.

Crossover and mutation include the vital operation utilized in genetic algorithm. The genetic algorithm (GA) is significantly utilized in dealing with the practical optimization issues. GA has the capability of handling both the discrete and continuous variables, which is suited for the purpose of handling the parallel computing and fitness environments that are extremely complex. Some researchers have designated that the genetic algorithms have a better adaptability to effectively deal with the various models instead of ancient mathematics programming techniques. In this

M. Jha (✉)
Citicorp Services India Limited, Pune, India

S. Singhal (✉)
School of Computing and Information Technology, Manipal University Jaipur, Rajasthan, India
e-mail: sunita.singhal@jaipur.manipal.edu

© Springer Nature Switzerland AG 2020

A. Haldorai et al. (eds.), *Business Intelligence for Enterprise Internet of Things*,
EAI/Springer Innovations in Communication and Computing,
https://doi.org/10.1007/978-3-030-44407-5_11

chapter, the main aim is to enhance the status of task scheduling based on the comparison of the proposed genetic algorithm with the present techniques to obtain the accuracy, fitness value, and runtime and know the turnaround time of the task allocated by various task schedulers like FCFS, RR, etc.

11.2 Proposed System

In the proposed system, we have purposed a genetic algorithm to calculate and measure the fitness value from the assigned tasks in the scheduler. The tasks schedulers could be in the form of FCFS, RR, etc. By using GA we will come to know the minimum turnaround time or execution time required for processing the tasks as shown in Fig. 11.1.

11.2.1 General Arrangement of a Genetic Algorithm

Generally, GA includes five elements:

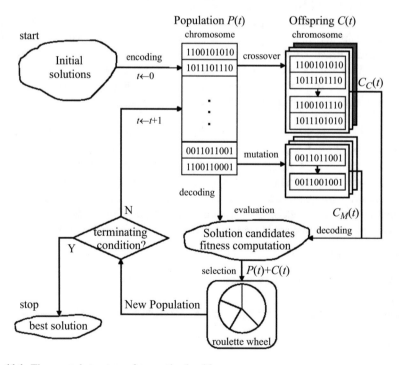

Fig. 11.1 The general structure of a genetic algorithm

1. A genetic account of probable resolutions to the issue
2. A technique of formulating a population (a preliminary assortment of possible solutions)
3. Testing functions in terms of fitness ranking solutions
4. Genetic operators altering offspring's genetic composition (crossover, mutation, selection, etc.)
5. The constraint values utilized by the genetic algorithm (population size, genetic operator probabilities, etc.)

11.2.2 Algorithm Techniques

The proposed algorithm technique for task scheduling is genetic algorithm as given below:

Algorithm: Genetic Algorithm

```
1  START
2  Generate the initial population
3  Compute fitness
4  REPEAT
5  Selection
6  Crossover
7  Mutation
8  Compute fitness
9  UNTIL secluding has converged
10 STOP
```

The given algorithm shows the execution process of the tasks allocated to the CPU. The GA includes primarily the initialization, discovery, crossover, and mutation phases. Initially all the tasks are given for processing. A set of parameters (variables) known as genes describe a person. The fitness function determines how competitive a person is (an individual's ability to compete with other people). The probability that an individual will be selected for processing is based on its fitness score. The whole process is repeated until the fit individuals are not obtained. Next step is for fitness score. The fitness element refers to the fitness of someone's state (an individual's ability to compete with other folks). It gives a fitness score to each individual. Next step is selection. Fitness-related selection, also known as selection of roulette wheels, is a genetic operator used in genetic algorithms to select potentially useful recombination solutions. In fitness-related selection, as in all selection strategies, the fitness function assigns fitness to possible solutions. A crossover is a simple operation wherein a pair of strings randomly swaps their substrings with each other. This operator's role in GA is more critical. It implements the principle of evolution. Next step is mutation. An operation of mutation is formulated to diminish the processors' idle time that is spent waiting for the other processors' information. Again the fitness of the process is computed for the selection procedure until the scheduling process is converged.

11.3   Literature Review

Shigeyoshi Tsutsui et al. [1] projected simplex crossover (SPX), a new multi-parent recombination operator for real-coded GAs. The SPX is considered as a simplified crossover operator, which utilizes the search space property of the simplex. The SPX is characterized with a balance between the development and discovery and is distinct in the production of offspring from coordinate systems. The results of the experiment showed high performance with multimodality and/or epistasis on test functions.

Isao Ono et al. [2] proposed a novel actual-coded genetic algorithm that utilizes the unimodal normal distribution crossover (UNDX) improved by the uniform crossover (UX). The UNDX is characterized by an advantage that is not found in other crossover operators. As such, this is effective for the process of optimizing the functions in operators with firm epistasis. UNDX is assured with the most crossover operators; hence, it is vital for optimizing various functions in the parameters with firm epistasis.

Zhu Can et al. [3] applied the genetic algorithms with actual codes to resolve issues meant to find the novel individual promised by the crossover operators of the seeds in various localized minimal points disrupting schemata of parents. This was responsible in acting as mutation operators. With reference to this, the idea is separated from the optimal sub-population and proposed by the population division centered on the Euclidean distances between the present optimal individuals. Kalyanmoy Deb et al. [4] recommended the standardized parent-centered recombination operators and the steady-state, scalable, elite-conserving, and rapid population shift models. The performance of the G3 system, including the PCX operator, is evaluated on three extensively utilized test issues as contrasted with various classical and evolutionary optimization algorithms, which include other actual-parameter GAs composed of uniform modal normal distribution crossover (UNDX) and simplex crossover (SPX) operators and the related self-adaptive evolutionary algorithm. The technique of covariance matrix adaptation evolution (CMA-ES), the methodology of differential evolution, and the quasi-Newton method are some of the initiatives. It was noticed in the analysis that the recommended algorithm performed an effective, reliable, and consistent approach. Saber M. Elsayed et al. [5] presented an objective meant to enhance the evaluation of the genetic algorithm through the process of initiating novel crossover with random operator targeted at replacing mutation. The projected crossover uses three parents vital for generating three new offspring relevant for exploitation, whereas a third offspring is purposed to enhance exploration. The randomized operator is helpful to facilitate an escape and premature convergence.

Can et al. [6] analyzed the unfair use of fair selection laws to overcome genetic algorithm restricted optimization. A form of population framework with the ideology of Pareto dominance, including the sequence of user factory designs. As a result, this balances the feasible limits and regions along an optimal search path of the resolution that makes the genetic algorithm of constrained optimization on all sides of the region feasible to consider the optimal resolution. There is a relation between the feasible part in demes and the revolutionary generation. It considers both the algorithmic search quality and efficiency optimization. The example of computational research and engineering has shown that the improved genetic algorithm of constrained optimization includes a simplified algorithm structure with a significant quality of resolutions. Tetsuyuki Takahama et al. [7] recommended a constrained DE with the archive and gradient-based mutation (εDEag) to enhance the stability, usability, and efficiency of the εDEg.

K. Sunitha et al. [8] designed an algorithm to schedule the DAG tasks on heterogeneous processors in such a way that minimizes the total completion time (makespan). Makespan is a measure of the throughput of the heterogeneous computing system (execution time + waiting time or idle time). Comparing makespan for different number of processors, number of tasks, population size, and number of generations to be done. Yan Kang et al. [9] have shown the use of GAs in the process of dealing with task scheduling issues through various means. The two vital techniques appear to be the methodologies that apply the GA in the linking of other list scheduling methods and techniques that apply the GA to produce the real assignment of transforming tasks to processors.

11.3.1   Expected Results

This section contrasts simulation results of our proposed algorithm with the results of several other algorithms. Table 11.1 lists the parameters of our proposed algorithm and the other algorithms used in the performance analysis. Table 11.2 indicates the detailed specifications.

By studying various techniques in literature survey, we will improve the methods and techniques in Figs. 11.2, 11.3, and 11.4. We also applied our algorithm to a task scheduling problem [10]. Task scheduling's primary goal is to schedule tasks on processors and reduce the make-up of the program, i.e., the accomplishment period of the last task compared to the start time of the first task. The output of the problem is the assignment of tasks to processors.

Table 11.1 Comparison of all algorithms along with their parameters and values

Algorithm	Parameter	Value
GGA	P-CROSSOVER	0.8
	P-MUTATION	0.8
	NUMBER OF GENERATION	VARIABLE
SCGA	P-CROSSOVER	0.8
	P-MUTATION	0.8
	NUMBER OF GENERATION	VARIABLE
SSGA	P-CROSSOVER	0.8
	P-MUTATION	0.8
	NUMBER OF GENERATION	VARIABLE
GA-RA	P-CROSSOVER	0.8
	P-MUTATION	0.8
	NUMBER OF GENERATION	VARIABLE
CLOUDSIM	MIPS(VM)	20(incremented by 10)
	CLOUDLET LENGTH	4000(incremented by 50)
	SCHEDULING POLICY	VARIED

Table 11.2 Algorithm's runtime and objective value for OneMax function

| | | OneMax | | |
| | gGA | | GA-RA | |
Iterations	Runtime	Value	Runtime	Value
5000	158	315	332	512
10000	372	332	550	512
15000	563	350	808	512

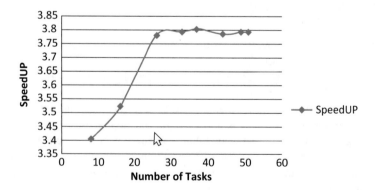

Fig. 11.2 Speedup vs. number of tasks

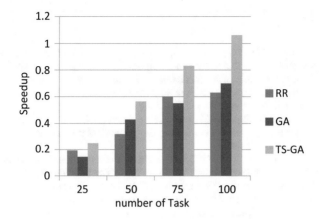

Fig. 11.3 The comparison speedup of three algorithms RR, GA, and TS-GA

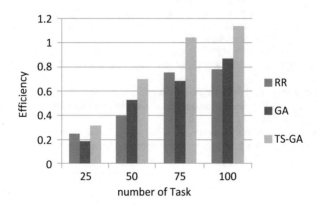

Fig. 11.4 The comparison efficiency of three algorithms RR, GA, and TS-GA

11.4   Summary

GA is a heuristic remedy search method motivated by the physical evolution. It belongs to the class of flexible and robust approaches. GA can be used in a wide range of optimization and learning issues. This is one of the major reasons for its enormous popularity. It is certainly suited to issues whereby traditional optimization systems break down due to irregular structure of search spaces or due to searches becoming computationally intractable. In this research, we have proposed a genetic algorithm for solving optimization problems. We also tested the algorithm on task scheduling problem and found it efficient in reducing the makespan time. The experimental examination presented that the algorithm congregates quicker than its counterparts to the optimal solution. Also, the results produced were effective based on the projected value, thus exhibiting a superior performance.

References

1. Shigeyoshi Tsutsui, Masayuki Yamamura, & Higuchi, T.. (1999). Multi-parent recombination with simplex crossover in real coded genetic algorithms. In *IEEE* (pp 657–664).
2. Ono, I., Kita, H., & Kobayashi, S. (1999). A Robust real-coded genetic algorithm using unimodal normal distribution crossover augmented by uniform crossover: Effects of self-adaptation of crossover probabilities. *IEEE, 1*, 496–503.
3. Zhu Can, & Liang Xi-Ming. (2009). Improved genetic algorithms to solving constrained optimization problems, international conference on computational intelligence and natural computing. In *IEEE* (pp 486–489).
4. Deb, K., Anand, A., & Joshi, D. (2002, December). A computationally efficient evolutionary algorithm for real-parameter optimization, evolutionary computation. *IEEE, 10*(4), 371–395.
5. Elsayed, S. M., Sarker, R. A., & Essam, D. L. (2011) GA with a new multi-parent crossover for constrained optimization. In *2011 IEEE Congress of Evolutionary Computation (CEC)* (pp. 857–864). New Orleans.
6. Can, Z., Xi-Ming, L., & Shu-renhu, Z. (2009). Improved genetic algorithms to solving constrained optimization problems. In *2009 International conference on computational intelligence and natural computing* (pp. 486–489). IEEE.
7. Takahama, T., & Sakai, S. (2010). Constrained optimization by the ε constrained differential evolution with an archive and gradient-based mutation. In *Congress on Evolutionary Computation* (pp. 1–9). Barcelona: IEEE.
8. Sunita, K., & Sudha, P. V. (2013). An efficient task scheduling in distributed computing systems by improved genetic algorithm. *International Journal of Communication Network Security, IEEE, 2*, 24–30.
9. Kang, Y., & Zhang, Z. (2011). An activity-based genetic algorithm approach to multiprocessor scheduling. In *2011 Seventh international conference on natural computation* (pp. 1048–1052). Shanghai: IEEE.
10. Anandakumar, H., & Umamaheswari, K. (2017). Supervised machine learning techniques in cognitive radio networks during cooperative spectrum handovers. *Cluster Computing, 20*(2), 1505–1515.

Chapter 12
An Algorithmic Approach to System Identification in the Delta Domain Using FAdFPA Algorithm

Souvik Ganguli ⓘ, **Gagandeep Kaur, Prasanta Sarkar, and S. Suman Rajest**

12.1 Introduction

System identification relates to the field of modeling dynamic systems from experimental data [1]. This process requires a huge amount of information regarding its input/output which may not always be available. In the classical methods, a model structure is first selected, and the unknown model parameters are identified by minimizing an objective function. However, conventional techniques are typically based on gradient descent techniques [2].

Metaheuristic algorithms and their hybridizations have already made inroads in identification and control literature [3, 4]. Wiener and Hammerstein systems are two commonly used prototypes used for the identification of systems [5]. Parameter assessment of these systems is conducted in the discrete time based on the application of the time domain shift operator and the complex domain z-transform using various soft computing techniques [6–10]. A large volume of works also exists in the continuous-time system using classical techniques [2]. Hammerstein and Wiener model identification in a continuous domain is based on the application of the metaheuristic method approaches which are rarely being investigated.

There are several methods built on discrete time systems using the ability of computers in system identification and control. Almost contemporaneously, there

S. Ganguli (✉) · G. Kaur
Thapar Institute of Engineering & Technology, Patiala, Punjab, India
e-mail: souvik.ganguli@thapar.edu; gagandeep@thapar.edu

P. Sarkar
National Institute of Technical Teachers' Training & Research,
Bidhannagar, West Bengal, India
e-mail: psarkar@nitttrkol.ac.in

S. S. Rajest
Vels Institute of Science, Technology & Advanced Studies, Chennai, Tamil Nadu, India

© Springer Nature Switzerland AG 2020　　　　　　　　　　　　　　　203
A. Haldorai et al. (eds.), *Business Intelligence for Enterprise Internet of Things*,
EAI/Springer Innovations in Communication and Computing,
https://doi.org/10.1007/978-3-030-44407-5_12

has been an analogous endeavor in evolving methods in systems theory because the physical signals are progressively by default. Modeling, control, and identification using the delta operator is continuous based on the application of the delta operator as comprehensive methodology whereby the systems and signals are structured in the discrete domain and leads to convergence to its corresponding continuous-time signals and systems at a high sampling frequency therefore unifying both the continuous- and discrete-time systems and signals [11].

Although, in the discrete-time domain, Hammerstein and Wiener model identification with metaheuristic approaches is absolutely familiar, similar analyses for continuous-time systems are rarely reported. Therefore, system classification with hybrid metaheuristic techniques can be considered for unification of discrete and continuous systems using the delta operator's properties. In order to estimate the unknown Hammerstein and Wiener model parameters in the delta domain, a hybrid algorithm, namely, FAdFPA, proposed by Ganguli et al. [12] was used. The remainder of the chapter is constituted as follows. Section 2 discusses the problem of identification in the delta domain. Section 3 gives an overview of the FAdFPA algorithm. Section 4 highlights the results, while Sect. 5 infers the paper.

12.2 Statement of the Problem

Hammerstein and Wiener models are two common models used in dynamic system identification. Ganguli et al. [13] formulated the simulation of the delta operator to estimate the unknown structure and parameters of polynomial nonlinearity using GWOCFA algorithm. Interested readers may go through their modeling equations that this chapter avoids. The authors of this chapter focused on developing a new heuristic approach to parameter calculation by minimizing the error resulting from the difference between real and expected values as described in Eq. 12.1.

$$J = \frac{1}{N}\sum_{k=1}^{N}\left[y(k) - \hat{y}(k)\right]^2 \tag{12.1}$$

where $y(k)$ and $\hat{y}(k)$ represent the actual and expected model performance responses. Therefore, in the next section, the proposed hybrid technique used to identify parameters in the delta domain is deliberated.

12.2.1 FAdFPA Algorithm

Previously, the authors developed a hybrid algorithm called FAdFPA to solve unconstrained problems of optimization [12]. By using FA as a global optimizer, the balance for exploration was achieved, whereas FPA was used to perform exploitation. To improve upon the local search abilities of FPA, its switch probability formulation has been made adaptive by the formula in Eq. 12.2

> Step1: Ignite the system with PRBS sequence adulterated with white noise.
>
> Step2: Yield random solutions for parameters of the linear and nonlinear parts in the specified search domain.
>
> Step3: Calculate the fitness function J for all possible solutions produced in Step 2.
>
> Step4: Using FAdFPA algorithm find the best possible solution and its position.
>
> Step5: Compare the best position with the previous best position.
>
> Step6: Update position for the new fitness function.
>
> Step7: Repeat Step 4 until the maximum iteration or best fitness value is produced.

Fig. 12.1 Steps for identification algorithm using FAdFPA

$$p = p_{max} - (p_{max} - p_{min}) \times (t / T) \tag{12.2}$$

Here, p_{max} and p_{min} are two user-defined parameters, considered as 0.9 and 0.4, respectively, in the experiments as reported in the literature. "T" denotes the maximum number of iterations while "t" is the present iteration. The steps for system identification algorithms using the abovementioned hybrid technique are shown in Fig. 12.1 listed below.

Step 1: Ignite the system with PRBS sequence adulterated with white noise.
Step 2: Yield random solutions for parameters of the linear and nonlinear parts in the specified search domain.
Step 3: Calculate the fitness function J for all possible solutions produced in Step 2.
Step 4: Using FAdFPA algorithm, find the best possible solution and its position.
Step 5: Compare the best position with the previous best position.
Step 6: Update position for the new fitness function.
Step 7: Repeat Step 4 until the maximum iteration or best fitness value is produced.

12.3   Results and Discussions

The transfer function plant model [2] in the continuous time is represented as

$$G(\rho) = \frac{b_0\rho + b_1}{\rho^4 + a_1\rho^3 + a_2\rho^2 + a_3\rho + a_4} = \frac{-\rho + 5}{\rho^4 + 23\rho^3 + 185\rho^2 + 800\rho + 2500} \tag{12.3}$$

The corresponding delta model at 100 Hz sampling frequency is given by

$$G(\delta) = \frac{b_0\delta + b_1}{\delta^4 + a_1\delta^3 + a_2\delta^2 + a_3\delta + a_4} = \frac{-0.8\delta + 4.5}{\delta^4 + 22\delta^3 + 180\delta^2 + 760\delta + 2200} \tag{12.4}$$

Two separate continuous nonlinearities, viz.,

$$y(k) = c_1 x(k) + c_2 x^2(k) + c_3 x^3(k) = x(k) + 0.5x^2(k) + 0.25x^3(k) \quad (12.5)$$

and

$$y(k) = \frac{x(k)}{\sqrt{c_1 + c_2 x^2(k)}} = \frac{x(k)}{\sqrt{0.10 + 0.90x^2(k)}} \quad (12.6)$$

are taken up, respectively, for identifying Wiener and Hammerstein model parameters in the delta domain. For the PRBS input signal, a sample size of 255 was considered. The input signal has been polluted with an unbiased signal, viz., white Gaussian noise of SNR 50 dB. For each of the experiments, a population segment of about 20 and a maximum number of iterations of a hundred are considered. The parent algorithms FA and FPA and the latest heuristics such as MFO, MVO, SCA, and SSA have been compared. For all the algorithms listed, standard parameter values are considered for comparison with the hybrid technique.

As metaheuristic algorithms are fundamentally stochastic, they have to be executed multiple times to generate meaningful numerical measures. Hence 20 test runs are executed for every algorithm to get significant statistical results. Wilcoxon rank-sum test [14] is also performed to validate the findings. The identified and actual parameters for the Hammerstein and Wiener frameworks in the delta domains are indicated in Tables 12.1 and 12.2, respectively. The results are compared with sufficient number of heuristic techniques reported in the literature. The closer parameter estimates to the actual values are marked with the help of bold letters.

From Tables 12.1 and 12.2, it is quite fair to say that the hybrid technique proves better than the several metaheuristic algorithms in the delta domain. Table 12.3 also provides the statistical measurements of the Wiener and Hammerstein model parameters in the delta domain. The best results obtained in each column of the table are highlighted with bold letters.

The hybrid method yields the minimum fitness function as compared to other algorithms. Since standard deviation with the hybrid approach turns out to be the least, it can be inferred that the algorithm is more robust than those considered for comparison. Moreover, to verify the validity of the results obtained, few additional statistical tests need to be done, verifying that the results did not happen all of a sudden. The non-parametric Wilcoxon rank-sum analysis is therefore conducted to verify the significance of the obtained results, and the measured p-values are cited as relevant metrics. A few selected p-values are reported in Table 12.4.

Table 12.4 suggests that the results of the hybrid method are meaningful for all other algorithms. The convergence curve is drawn in Fig. 12.2 for the test system used for Wiener model identification in the delta domain. The y-axis of the plot represents the normalized fitness function while the x-axis depicts the number of iterations. Further, the parent algorithms FA and FPA as well as other popular algo-

Table 12.1 Comparison of Wiener system model parameters

Types of values	Algorithms	b_0	b_1	a_1	a_2	a_3	a_4	c_1	c_2	c_3
Actual		−0.8	4.5	22	180	760	2200	1	0.5	0.25
estimated	FAdFPA	**−0.7993**	**4.4856**	**21.9989**	**179.8492**	**759.8219**	**2206.2981**	**1.0018**	**0.5101**	**0.2494**
	FA	−0.7572	4.3832	21.7958	172.1841	763.0585	2258.0174	1.0442	0.4592	0.2620
	FPA	−0.8344	4.5947	22.0104	175.9721	766.0472	2370.6045	1.0576	0.5189	0.2565
	MFO	−0.7265	4.7379	23.1151	169.4181	734.5583	2251.8777	0.9507	0.4694	0.2538
	MVO	−0.7271	4.7374	23.1263	169.4236	734.5486	2251.898	0.9422	0.4492	0.3143
	SCA	−0.7753	4.2673	21.7239	164.1472	741.4551	2312.0279	0.9608	0.4675	0.2636
	SSA	−0.8429	4.4526	22.2516	172.5593	699.8514	2309.6354	1.0326	0.5385	0.2485

Table 12.2 Comparison of Hammerstein system model parameters

Types of values	Algorithms	c_1	c_2	b_0	b_1	a_1	a_2	a_3	a_4
Actual		0.1	0.9	−0.8	4.5	22	180	760	2200
estimated	FAdFPA	**0.9993**	**0.9023**	**−0.7943**	**4.4928**	**22.4527**	**179.8753**	**759.6725**	**2196.7754**
	FA	0.1009	0.9099	−0.8623	4.7801	20.6720	188.4102	763.1495	2031.5233
	FPA	0.0909	0.8992	−0.8045	4.4909	23.1408	164.1544	825.5786	2031.3154
	MFO	0.0992	0.8659	−0.7829	4.5010	21.4638	188.7328	804.1881	2237.4300
	MVO	0.0994	0.8758	−0.7927	4.5210	21.4936	188.7438	805.1892	2237.4538
	SCA	0.0983	0.9848	−0.7929	4.6235	20.1478	165.2356	793.3553	2275.4090
	SSA	0.1063	0.9206	−0.8153	4.4029	20.0600	166.9278	808.3730	2076.1786

Table 12.3 Statistical analysis of the fitness function

Test systems	Test methods	Minimum	Maximum	Mean	Standard deviation
Wiener model	FAdFPA	**0.0016**	**0.0017**	**0.0016**	**3.6635e-05**
	FA	0.0018	0.0020	0.0019	5.8361e-05
	FPA	0.0019	0.0021	0.0020	7.4325e-05
	MFO	0.0018	0.0020	0.0019	5.8296e-05
	MVO	0.0018	0.0021	0.0019	5.8356e-05
	SCA	0.0018	0.0019	0.0019	6.7876e-05
	SSA	0.0018	0.0020	0.0019	5.5030e-05
Hammerstein model	FAdFPA	**0.0183**	**0.0186**	**0.0184**	**4.9132e-05**
	FA	0.0190	0.0196	0.0193	5.9423e-05
	FPA	0.0190	0.0196	0.0193	7.7482e-05
	MFO	0.0191	0.0197	0.0193	5.9376e-05
	MVO	0.0191	0.0198	0.0194	5.9772e-05
	SCA	0.0191	0.0195	0.0193	7.8523e-05
	SSA	0.0190	0.0195	0.0193	9.7813e-05

Table 12.4 Selected significant p-values using non-parametric test

Test systems	Proposed method	FA	FPA	MFO	SCA
Wiener system	FAdFPA	7.9026E-05	5.6379E-05	1.2893E-06	2.9131E-06
Hammerstein system	FAdFPA	4.9017E-05	5.3492E-05	1.1300E-02	1.2000E-02

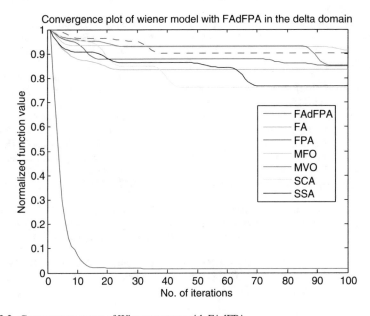

Fig. 12.2 Convergence curve of Wiener system with FAdFPA

rithms like MFO, MVO, SCA, and SSA are used for comparison with the proposed method.

It is obvious from Fig. 12.2 that the planned process congregates at a faster rate as compared to the other algorithms considered. In addition, the proposed method provides better accuracy in terms of the fitness function value.

12.4 Summary

The hybrid topology assimilating FA and FPA is used to recognize Hammerstein and Wiener model parameters in the delta domain by minimizing the fitness function. Delta operator modeling unifies continuous and discrete-delta domain results. The hybrid technique's fitness value not only outperforms that obtained by some latest metaheuristic algorithms but also the heuristics that it constitutes for the test systems considered. Also, the Wilcoxon test verifies the significance of the hybrid approach results. Similar to other algorithms, the hybrid approach shows superior convergence characteristic in the delta domain. Therefore, in the future, the algorithm can also be used to estimate fractional-order systems in the delta domain.

References

1. CLjung, L. (1987). *System identification: Theory for the user*. Upper Saddle River: Prentice-hall.
2. Sinha, N. K., & Rao, G. P. (2012). *Identification of continuous-time systems: Methodology and computer implementation* (Vol. 7). Dordrecht: Springer Science & Business Media.
3. Bhushan, B., & Singh, M. (2011). Adaptive control of DC motor using bacterial foraging algorithm. *Applied Soft Computing, 11*(8), 4913–4920.
4. Xu, X., Wang, F., &Qian, F.(2011). Study on method of nonlinear system identification. In *International Conference on Intelligent Computation Technology and Automation (ICICTA)* (Vol. 1, pp. 944–947).
5. Al-Duwaish, H. N. (2000). A genetic approach to the identification of linear dynamical systems with static nonlinearities. *International Journal of Systems Science, 31*(3), 307–313.
6. Pal, P. S., Kar, R., Mandal, D., & Ghoshal, S. P. (2016). Identification of NARMAX Hammerstein models with performance assessment using brain storm optimization algorithm. *International Journal of Adaptive Control and Signal Processing, 30*(7), 1043–1070.
7. Sun, J., & Liu, X. (2013). A novel APSO-aided maximum likelihood identification method for Hammerstein systems. *Nonlinear Dynamics, 73*(1–2), 449–462.
8. Raja, M. A. Z., Shah, A. A., Mehmood, A., Chaudhary, N. I., & Aslam, M. S. (2016). Bio-inspired computational heuristics for parameter estimation of nonlinear Hammerstein controlled autoregressive system. *Neural Computing and Applications, 29*(12), 1455–1474. (2018).
9. Cuevas, E., Díaz, P., Avalos, O., Zaldívar, D., & Pérez-Cisneros, M. (2018). Nonlinear system identification based on ANFIS-Hammerstein model using gravitational search algorithm. *Applied Intelligence, 48*(1), 182–203.

10. Garnier, H., Mensler, M., & Richard, A. (2003). Continuous-time model identification from sampled data: Implementation issues and performance evaluation. *International Journal of Control, 76*(13), 1337–1357.
11. Middleton, R. H., & Goodwin, G. C. (1990). *Digital control and estimation: A unified approach (prentice hall information and system sciences series)*. Prentice Hall: Englewood Cliffs, NJ.
12. Ganguli, S., Kaur, G., & Sarkar, P. (2016). A novel hybrid metaheuristic algorithm to solve unconstrained optimization problems. *International Journal of Advanced Computational Engineering and Networking, 4*(5), 40–43.
13. Ganguli, S., Kaur, G., & Sarkar, P. (2019). A hybrid intelligent technique for model order reduction in the delta domain: A unified approach. *Soft Computing, 23*(13), 4801–4814.
14. Wilcoxon, F., Katti, S. K., & Wilcox, R. A. (1970). Critical values and probability levels for the Wilcoxon rank sum test and the Wilcoxon signed rank test. *Selected tables in mathematical statistics, 1*, 171–259.

Chapter 13
An IoT-Based Controller Realization for PV System Monitoring and Control

Jyoti Gupta, Manish Kumar Singla, Parag Nijhawan, Souvik Ganguli (iD), **and S. Suman Rajest**

13.1 Introduction

In the recent era, the rapid increase in the consumption of electricity and issues regarding environmental concerns are the main reasons behind the development of renewable energy sources (RESs). The electricity generation from RES has been derived for its economic benefits and reliability [1]. Micro-grid includes distributed generation resources such as photovoltaic, wind, diesel generators, etc. The implementation of micro-grid in the distribution network was introduced in the early years of 2000 and encouraged by various agencies and utilities. Micro-grid has increased the operation of power electronics converter in the power system for efficient and effective power quality [2]. The micro-grid in a distribution system can be installed near the substation or at the end of the feeder. Micro-grid has an essential and useful feature, i.e., it can be operated in two different modes, islanded and grid-connected mode, to give essential support during the time of grid failure or maintenance of grid system by supplying constant power. Due to an increase in the interconnection of micro-grid in the distribution system, it raises some problems that include fluctuation in the voltage, steady-state over-voltage, increases in the system loss, and issues related to the voltage regulation devices and protection; therefore, appropriate allocation of micro-grid is desirable.

There are various types of micro-grids, namely, DC micro-grid, AC micro-grid, and hybrid micro-grid, which combines both AC and DC micro-grids [3]. Solar energy systems are dependent on two factors: temperature and irradiance. The max-

J. Gupta · M. K. Singla · P. Nijhawan · S. Ganguli (✉)
Thapar Institute of Engineering and Technology, Patiala, India
e-mail: jgupta_phd19@thapar.edu; souvik.ganguli@thapar.edu

S. S. Rajest
Vels Institute of Science, Technology & Advanced Studies, Chennai, India

© Springer Nature Switzerland AG 2020   213
A. Haldorai et al. (eds.), *Business Intelligence for Enterprise Internet of Things*,
EAI/Springer Innovations in Communication and Computing,
https://doi.org/10.1007/978-3-030-44407-5_13

imum power point (MPP) is the point where maximum power output is produced by PV array. The control scheme is developed using a neural network. A neural network (NN) is designed to provide a gate pulse signal to the inverter circuit to improve the performance of the system. The artificial neural network (ANN) is based on a machine learning approach and contains the number of artificial neurons to perform the specific task in the system [4]. To investigate its impact in different conditions, the inverter control scheme is used to improve the efficiency of the power transfer between the PV system and the grid. An IEEE 13-node test feeder system is considered as a grid system and is commonly used to test features of a standard distribution network of a power system, operating at 4.16 kV.

The main problem that arises in monitoring the output power of the photovoltaic system is the accuracy and time duration for the detection of the fault and the appropriate solution to it. The best approach for dealing with such kind of issues is the Internet of Things (IoT) [5, 6]. It is the new concept which has emerged recently and has gained a lot of attention in a few years. It can be generally explained as an information sharing environment where elements of the system are attached to a wireless and wired network. These days, this concept is not only applicable in the field of electronics but also in the area of home appliances, smart cars, industrial security, etc.

For the individuals and companies associated with solar panels, the IoT makes it possible to increase the MPPT reliability and performance of the system [7, 8]. For the controlling or monitoring of the system, the proposed method is discussed in the chapter.

13.2   Micro-grid Model Description

An IEEE 13-node test feeder system is considered as a grid system and is commonly used to test features of a standard distribution network of a power system, operating at 4.16 kV.

13.2.1   Photovoltaic Module

The photovoltaic system operates on the principle of the photovoltaic effect, that is, when sunlight is radiated upon the semiconductor diode, there is a movement of electrons from P-type to N-type side of the semiconductor which produces the current in the system [10].

The photovoltaic module is simulated in MATLAB/Simulink using a photovoltaic array. To generate power of 341.65 kW, the SunPower SPR-445NX-WHT-D model is used, selected from the module block in the PV array block. Also, there are 8 series-connected modules and 96 parallel-connected modules per string. The module parameters and data are shown in Table 13.1. These values are calculated standard temperature and irradiance, i.e., 25°C and 1000 W/m^2 irradiance.

Table 13.1 Model parameters
of PV array

Name of the parameters	Number/rating
Maximum power (watt)	444.86
Cell per module	128
Light generated current I_p(A)	6.2167
Diode saturation current I_D (A)	1.3552e-11
Shunt resistance R_{sh} (ohm)	508.2463
Series resistance R_{se} (ohm)	0.54861

13.2.2   DC-Link Capacitor and Inverter Circuit

A three-phase inverter circuit is connected across the DC link. DC-link capacitors are employed to stabilize the DC-link voltage of the grid-connected inverter. Due to temperature and irradiance variation to the total resistance within PV cells leading to non-linear output efficiency, capacitance is needed to allow electronic control methods to maintain maximum output power. The three-phase inverter converts the DC output voltage, i.e., 590V, from micro-grid into AC output, i.e., 240V.

13.2.3   RL Filter, Transformer, and Load

RL filter is connected at the three-phase inverter circuit output, to reduce the total harmonics distortion in the current. They prevent overcurrent condition in the system. A three-phase coupling transformer is between the micro-grid system and grid. Three-phase coupling transformer, star to the delta, is connected to step-up the voltage from 240V to 4160V at 50Hz frequency. It has a winding resistance of 0.01 per unit. The load is the distributed load of IEEE 13-node feeder systems.

13.2.4   Point of Common Coupling and Grid

Point of common coupling is described by using circuit breakers between PV system and grid system; it plays a vital role in the proposed method by isolating the PV system from grid system, i.e., in the islanded mode at the time of grid blackout or when grid is under maintenance and to supply power to particular load in case of deficit energy produced by micro-grid. Grid is a standard IEEE 13-node test feeder system, as displayed in Fig. 13.1.

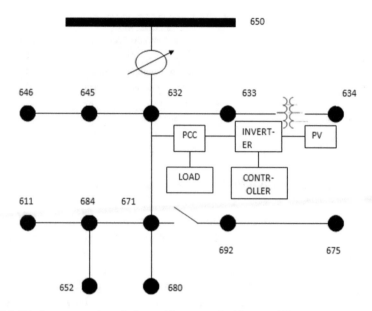

Fig. 13.1 Block representation of micro-grid-connected grid system [9]

13.3 Proposed Control Features

Micro-grid is mainly operated in two modes, i.e., islanded and grid-connected. The proper controlling action of micro-grid is an essential condition for steady and efficient operation in every mode. The controller performs the following functions:

- Control of the power flow between the grid system and micro-grid.
- Proper synchronization of micro-grid with the grid system.
- Proper regulation of load sharing between the grid system and micro-grid.
- Regulation of the frequency and voltage for both operating modes, namely, grid-connected and islanded mode.
- Re-optimization of the operation cost of the micro-grid.
- While operating during the switching modes, proper handling of transients and restoration of the desired condition are necessary.

13.3.1 MPPT Controller

Maximum power point tracking (MPPT) is the algorithm which extracts the maximum power from the photovoltaic system generated by photovoltaic cell or module at the specific environmental condition. The specific voltage at which solar device generates maximum power output is called maximum power point. There are various types of techniques used for maximum power point tracking, out of which perturb and observe algorithm is used in this work for determining the maximum power point of the photovoltaic module. This technique uses only one sensor to detect the maximum power point that is a voltage sensor. It also reduces the cost and complexity of the system. This technique is quite easy as it has the least time complexity [11, 12].

13.3.2 Inverter Controllers

13.3.2.1 PI Controller

The inverter controller operates on the double closed-loop current algorithm. There are two loops, outer and inner loop. The voltage control loop is referred to as the outer loop, and its primary duty is to maintain the DC voltage of PV array. A current control loop is referred to as the inner loop, and its primary duty is to keep the active and reactive current signal of the grid (I_d and I_q). For stabilizing the power factor at unity for efficient grid interconnection, the voltage control loop provides the output of I_d, which is the reference of current, while keeping I_q value regulated to minimum, i.e., zero, using a PI controller to minimize error. The inner control loop provides the voltage output V_d and V_q. The tracking error can be reduced, and tracking speed can be increased by regulating the PI controller in the inner current control loop. Forward-feed compensation output is included in the loop to decrease the disadvantages to the perturbation of the voltage of grid.

The phase-locked loop (PLL) algorithm is applied to regulate the values of the active power and reactive power injected into the IEEE 13-node feeder systems. The advantage of a phase-locked loop is that it helps in locking and synchronizing the frequency and phase angle output of the voltage with reference to the current of the grid using transformations. The output of the current control loop and voltage control loop is feed to the converter block whose function is to sense the vector values of current and voltage, respectively and convert theses values into the d-q reference frame as DC quantities. The output of the converter block is then feed to over modulation in order to increase the linear region of a three-phase PWM modulator by approximately 15%. The output is transferred to PWM three-level pulse generators to produce the input pulse to the inverter.

13.3.2.2 Artificial Neural Network Controller

An artificial neural network is a massively parallel distributed processor consisting of simple processing units, which has an inherent property to store knowledge of experiments and make it available for use. In two respects it resembles the brain as follows:

- The strengths of the interneuron connection, known as synaptic weights, are used to store the knowledge gained.
- The network acquires knowledge from its environment through a learning process.

- Mathematically, neuron modeling can be represented by the following Eqs. 13.1, 13.2 and 13.3:

$$u_k = \sum_{j=1}^{M} w_{kj} x_j \tag{13.1}$$

$$y_k = \varphi\left(u_k + b_k\right) \tag{13.2}$$

$$v_k = u_k + b_k \tag{13.3}$$

where x_j denotes the inputs, w_{kj} represents the synaptic weights of the neuron k, u_k is the combined output due to the inputs, b_k is the bias, $\varphi(.)$ is the activation function, and y_k is the output signal of the neuron. The use of bias has the effect of applying an affine transformation to the linear combiner output u_k in the modeling process.

Levenberg-Marquardt algorithm is used to train the neural network, which is the second-order optimization of the backpropagation algorithm. This algorithm typically requires less time and is robotic. This algorithm typically requires less time and is robust, which are the advantages of this algorithm over the Gauss-Newton algorithm and gradient descent method. The flow chart explaining the process of a neural network is shown in Fig. 13.2.

13.3.2.3 Monitoring System

Mainly two main MPPT designing variables, perturbation period (Tp) and perturbation magnitude (x), can be used as metrics. For the higher-resolution system, a smaller metric has greater acceptability [13–15]. The primary constraint of the mentioned system is the speed at which it receives and regulates the desired information. The constraints of the MPPT variable Tp are described in Eq. 13.4.

$$Tp >= \frac{-\ln\left(\dfrac{\in}{2}\right)}{\delta.\omega_n} \tag{13.4}$$

where w_n is the natural frequency of the defined inverter system, δ is the damping factor of the system, $\in$ is the controlling variable of the system, and mainly, 0.1 is the assigned value. The constraints of variable (x) are described in Eq. 13.5.

$$x > \frac{1}{\mu}.\sqrt{A.Kph.|G|.Tp} \tag{13.5}$$

Here, μ is the static gain of the inverter, Kph is the PV material constant related to the spectral-averaged responsiveness, G is the average of the irradiance slope, and A is the combination of dependent variable irradiance [16, 17]. Figure 13.3 represents the block diagram of the IoT-based MPPT system of the modules.

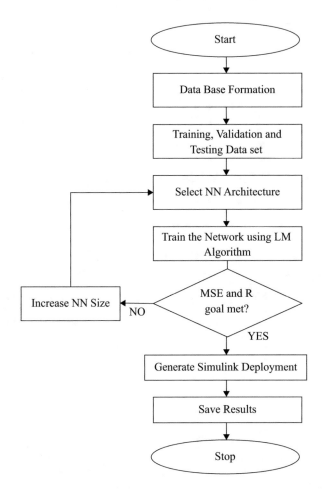

Fig. 13.2 Flow diagram of ANN

13.4 Results and Discussions

The system is simulated in Simulink for 0.75 s with the following parameters: frequency of 50Hz at standard irradiance and temperature and 590V. The allocation of PV system between the load bus 632 and 631 in the IEEE 13-node feeder system is decided by the calculation of the central position in the grid using line segment data.

The inverter circuit is controlled by using a standard PI controller and neural network. The power quality results using both the controllers are compared. The power quality results include power output, voltage deviation, total harmonic distortion, and phase angle deviation, which are observed as shown in Figs. 13.4 (a), 13.4(b), 13.5(a), 13.5(b), and 13.6, respectively.

Fig. 13.3 IoT-based
MPPT system [17]

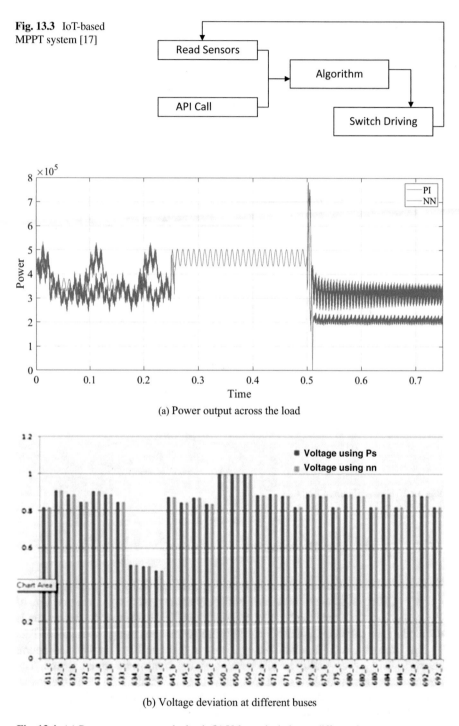

(a) Power output across the load

(b) Voltage deviation at different buses

Fig. 13.4 (**a**) Power output across the load. (**b**) Voltage deviation at different buses

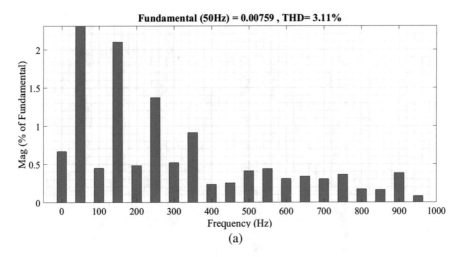

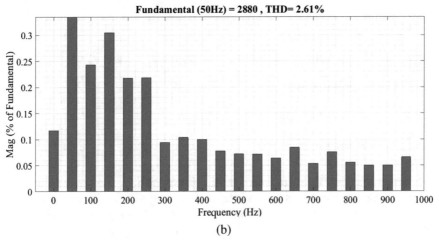

Fig. 13.5 (**a**) FFT using PI and (**b**) FFT using NN

The output graphs of the power quality are divided into three half-sections:

1. The first part is from 0 s to 0.25 s; at this time, system load is met by both PV system and grid, i.e., the system is operated at grid-connected PV mode.
2. The second part is from 0.25 s to 0.5 s; at this time, system load is met only by grid, i.e., PV output is not sufficient to meet the required load.
3. The third part is from 0.5 s to 0.75 s; at this time, system load is achieved only by PV system, i.e., PV output is sufficient to satisfy load, and the system is operated at islanded mode.

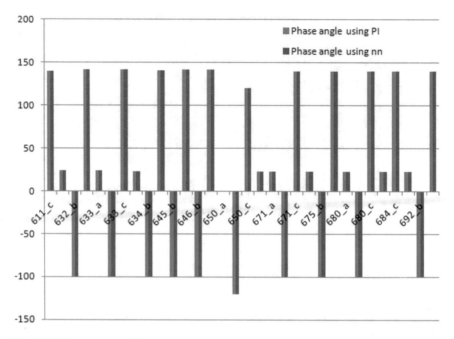

Fig. 13.6 Phase angle at different buses

13.5 Summary

It is presented that the interconnection of a photovoltaic system as a micro-grid improves the reliability of electric supply and power quality by regulating the IoT-based inverter controller systems. The IoT-based ANN inverter control scheme for the efficient and reliable power transfer between the grid system and micro-grid to satisfy the desired load is being demonstrated in this work. The IoT-based ANN-based controller response has also been compared with IoT-based PI controller, which justifies its technical feasibility.

References

1. Sadeghian, Hamidreza, Mir HadiAthari, & Zhifang Wang. (2017). Optimized solar photovoltaic generation in a real local distribution network. In *2017 IEEE Power & Energy Society Innovative Smart Grid Technologies Conference (ISGT)* (pp. 1–5). IEEE.
2. Lasseter, R. H. (2002). Micro-grids. In *2002 IEEE Power engineering society winter meeting. Conference Proceedings (Cat. No.02CH37309) 1* (pp. 305–308). IEEE.
3. Pham, D. H., Hunter, G., LiLi, & Zhu, J.. (2015). Advanced microgrid power control through grid-connected inverters. In *2015 IEEE PES Asia-Pacific Power and Energy Engineering Conference (APPEEC)* (pp. 1–6). IEEE.

4. Singh, K., Swathi, P., &Ugender Reddy, M.. (2014). Performance analysis of PV inverter in microgrid connected with PV system employing ANN control. In *2014 International Conference on Green Computing Communication and Electrical Engineering (ICGCCEE)* (pp. 1–6). IEEE.

5. Chen, X., Sun, L., Zhu, H., Zhen, Y., & Chen, H.. (2012). Application of internet of things in power-line monitoring. In *2012 International conference on cyber-enabled distributed computing and knowledge discovery* (pp. 423–426). IEEE.

6. Kang, B., Park, S., Lee, T., & Park, S.. (2015). IoT-based monitoring system using tri-level context making model for smart home services.In *2015 IEEE International Conference on Consumer Electronics (ICCE)* (pp. 198–199). IEEE.

7. Woyte, A., Richter, M., Moser, D., Mau, S., Reich, N., & Jahn, U.. (2013). Monitoring of photovoltaic systems: good practices and systematic analysis. In *Proceedings of the 28th European photovoltaic solar energy conference* (pp. 3686–3694).

8. Belghith, O. B., & Sbita, L.. (2014). Remote GSM module monitoring and photovoltaic system control.In *2014 First International Conference on Green Energy ICGE 2014* (pp. 188–192). IEEE.

9. Kersting, W. H.(2009). Distribution feeder voltage regulation control. In *2009 IEEE rural electric power conference* (pp. C1–C1). IEEE.

10. Villalva, M. G., Gazoli, J. R., & Filho, E. R. (2009). Comprehensive approach to modeling and simulation of photovoltaic arrays. *IEEE Transactions on power electronics, 24*(5), 1198–1208.

11. Chin, C. S., Tan, M. K., Neelakantan, P., Chua, B. L., & Teo, K. T. K. (2011). Optimization of partially shaded PV array using fuzzy MPPT. In *2011 IEEE colloquium on humanities, science and engineering* (pp. 481–486). Penang: IEEE.

12. Esram, T., & Chapman, P. L. (2007). Comparison of photovoltaic array maximum power point tracking techniques. *IEEE Transactions on Energy Conversion, 22*(2), 439–449.

13. De Brito, M. A. G., Galotto, L., Sampaio, L. P., de Guilherme, A. e. M., & Canesin, C. A. (2012). Evaluation of the main MPPT techniques for photovoltaic applications. *IEEE Transactions on Industrial Electronics, 60*(3), 1156–1167.

14. Esram, T., & Chapman, P. L. (2007). Comparison of photovoltaic array maximum power point tracking techniques. *IEEE Transactions on Energy Conversion, 22*(2), 439–449.

15. Beevi, S.Shanifa, J. M., & Vincent, G.. (2016). A high performance instantaneous resistance maximum power point tracking algorithm. In *20167th India International Conference on Power Electronics (IICPE)* (pp. 1–5). IEEE.

16. Ma, J., Man, K. L., Ting, T. O., Zhang, N., Guan, S. U., & Wong, P. W. H. (2014). Estimation and revision: A framework for maximum power point tracking on partially shaded photovoltaic arrays. In *2014 International symposium on computer, consumer and control* (pp. 162–165). Taichung: IEEE.

17. Bardwell, M., Wong, J., Zhang, S., & Musilek, P. (2018). Design considerations for IoT-based PV charge controllers. In *2018 IEEE world congress on services (SERVICES)* (pp. 59–60). San Francisco: IEEE.

Chapter 14
Development of an Efficient, Cheap, and Flexible IoT-Based Wind Turbine Emulator

Manish Kumar Singla, Jyoti Gupta, Parag Nijhawan, Souvik Ganguli (ID),
and S. Suman Rajest

14.1 Introduction

Energy is always the most crucial need for growth and development of a nation. Due to the rapid growth of the world population, changes in technology, and other political and economic scenarios, the energy demand is increased immensely. As of now, though more than 80% of the energy is generated by conventional sources of energy, it is gradually depleting. So, the development of an efficient and reliable renewable system becomes the prime motive [1]. Wind technology is rapidly improving in efficiency and scope. Energy researchers have high future goals for wind energy development. The main difficulty for the development of the wind energy system is the natural disruption of these resources. If units are too large to meet demand, severe imbalances can endanger system security. IoT has emerged as a smart and effective solution for monitoring the renewable energy system. Internet of Things (IoT) can be explained as a digital network that connects various elements of a particular system and regulates the data according to the situation using advanced embedded systems, including controllers, meters, and sensors. IoT brought the third revolution in the field of technology [2]. In the near future, IoT is expected to be widespread and will cover all aspects of human life, including the production and management of renewable energy.

An advanced version of IoT is introduced named as the Internet of Energy (IoE), which includes the arrangement of energy ecosystem and ICT. The implementation

M. K. Singla · J. Gupta · P. Nijhawan · S. Ganguli (✉)
Thapar Institute of Engineering and Technology, Patiala, India
e-mail: jgupta_phd19@thapar.edu; souvik.ganguli@thapar.edu

S. S. Rajest
Vels Institute of Science, Technology & Advanced Studies, Chennai, India

© Springer Nature Switzerland AG 2020   225
A. Haldorai et al. (eds.), *Business Intelligence for Enterprise Internet of Things*,
EAI/Springer Innovations in Communication and Computing,
https://doi.org/10.1007/978-3-030-44407-5_14

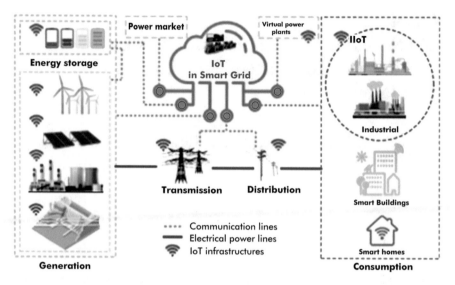

Fig. 14.1 Application of IoT in smart grid [4]

of IoT as a monitoring system in the smart grid is described in this study [3]. Nowadays, IoT is considered a vital part of the designing and execution of smart building schemes and smart cities, as shown in Figure 14.1. With the advancement in the field of IoT, a new theory is propounded defined as smart grid 2.0 which is also known as the second generation of smart grids. It refers to the improved model of the smart hybrid system with the IoT and is expected to be implemented in the next few years [4].

In the new concept (smart grid 2.0), the power exchange between the hybrid energy system and the power grid is regulated using the advanced-level smart electric metering system. The amount of energy exchange, the plug and play capability, and the vital information about the parameter between seller and buyer could be shared using informatics infrastructure. Plug and play feature indicates that a demand-side electricity source can inject power to the grid as easy as inserting a plug into an outlet, i.e., ability to inject even a very minimal power generation into the power grid like vehicle-to-grid (V2G) power transfer. Such types of consumers which have the ability to deliver power to another system are defined as prosumers [5]. Still, now there are no technologies that can disconnect and connect the distributed generation system with a grid at each desired moment. It is sensible to study wind energy conversion system (WECS) with respect to cybersecurity. The overall effect and significance of cybersecurity on the reliability of WECS have been studied [6, 7].

14.2   Wind Energy System with IoT

Wind energy is nowadays emerging as one of the most promising technologies. However, the output of the wind energy conversion system (WECS) is dependent on wind flow, which is sometimes erratic and at times unpredictable. By this time, developing a control scheme and ensuring a reliable WECS has been a significant area of focus. However, the efficacy of the control scheme can be assured by several experiments with an actual wind turbine (WT). The actual wind turbine is ultimately dependent on environmental conditions. Such dependency may cause indefinite delays. Moreover, several worst cases for which system needs to be tested may never occur, and thus performance and reliability of controllers for those cases are always questionable. Another way is to set up a WT in the laboratory. But it is difficult to set up a WT in the laboratory because of challenges like space and controlled environment, for example, wind tunnel.

A solution to these problems can be found in a wind turbine emulator. Wind turbine emulator mimics the behavior of actual wind turbine under controlled manner. Line diagram of wind turbine emulator and power conditioning unit is shown in Fig. 14.2. Essentially it simulates the same operating pattern at hardware level in real time similar to what an actual wind turbine does at given operating parameters of wind speed and pitch angle. The term emulation is coined for simulation practices which involve hardware platform. It provides a fast-configurable testing platform. Usually these are performed in real time. Emulator is similar to hardware-in-the-loop simulation concepts. Wind turbine emulator can find applications in numerous fields. It provides a flexible testing platform for the study of dynamic and steady-state behavior of wind turbine. Beginners can learn about power/wind speed, torque/turbine speed, and power/turbine speed characteristics of a wind turbine and can perform comparative studies about how changes in parameters of wind turbine affect the behavior. For other applications, this emulator can be coupled to the generator (induction, PMSG, DFIG) followed by power electronics in place of an actual wind turbine. So, researchers would not have to rely on environmental conditions which are appropriate for driving the wind turbine at some desired operating point. Since the operating point of wind emulator can be controlled, researchers can simulate all possible scenarios of operation of a wind turbine and accordingly can modify the power electronics and control algorithms. Consequently, it will improve the product quality and reliability. The operating point of wind emulator can be controlled using IoT. The block diagram of wind operating system with IoT system is shown in Figure 14.3.

IoT topology in combination with ICT infrastructure allows the wind power producers to regulate the output power at maximum efficiency and provides the accurate maintenance reminder of each component at regular interval in order to avoid any huge disaster. There are various algorithms which help in formulating the desired schedule for maintenance like machine learning, fuzzy logic, neural network, etc. For instance, on-time maintenance can reduce the index of levelized

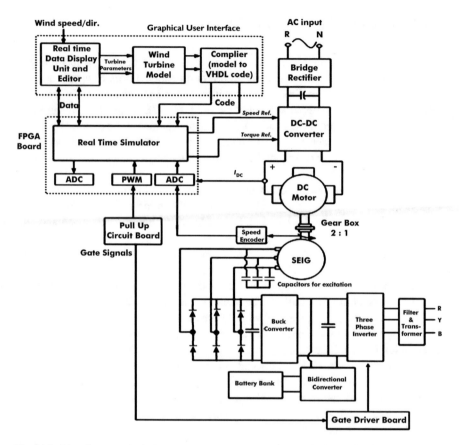

Fig. 14.2 Line diagram of wind turbine emulator and power conditioning unit

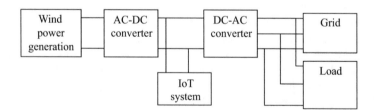

Fig. 14.3 Block representation of wind operating system with IoT system

energy costs (LCoE) for wind assets that denotes the net present value of the unit power cost over the lifetime of the turbines [8].

In digital communication system, the beneficiary data of each element of system is collected and processed using machine learning algorithm [9]. The main two problems related with the IoT system are the delay in information transfer especially in offshore wind farms and limitation on the bandwidth for exchanging information. Therefore, if essential information could be received and processed at a faster rate, then corrective measures for the system at the time of failure can be automated like shutting down of turbine system in case of turbulence. Therefore, we need the deployment of more IoT systems with improved algorithm for monitoring the wind power generation system [10].

14.3 Results

The wind emulator system developed is run at varying motor buck duties in order to obtain the effect on the various output parameters such as generator voltage, generator current, generator speed, battery voltage, DC link voltage, and motor armature current. The following results are also obtained using LabVIEW software. Their variation with time is shown in Table 14.1.

Thus, all the output parameters are obtained in the graph with respect to time which are shown in Figure 14.4 and are presented in two parts.

Speed is varying with change in the duty ratio while field and armature voltages remain invariant. The graph shown in Fig. 14.4 provides the variation of the motor output parameters versus the time. Thus, an efficient, economical, and flexible IoT-based wind turbine emulator is successfully developed in this chapter.

Table 14.1 Results of motor parameters

Motor buck duty (%)	Field voltage (V)	Armature voltage (V)	Speed (rpm)
2	230	229	60
4	230	229	74
6	230	229	96
8	230	229	116
10	230	229	131
12	230	229	146
14	230	229	164
16	230	229	180
18	230	229	199
20	230	229	215
22	230	229	229

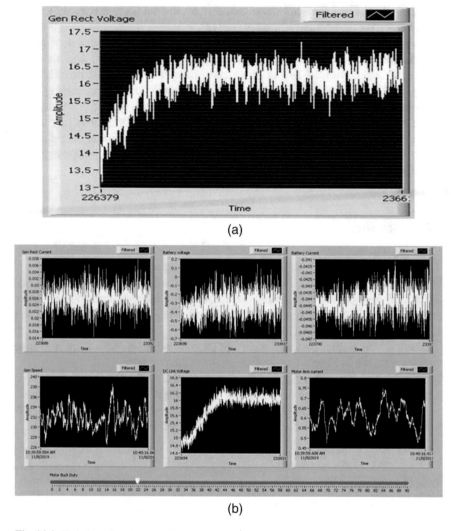

Fig. 14.4 Variation of motor output parameters with time

14.4 Summary

The proposal for a highly efficient, cost-effective, and flexible IoT-based wind turbine emulator is thus presented in this chapter. The monitoring can be performed using a web-based service which would receive the wind turbine data and convert it into useful information. The challenge however still lies with the feasibility of the amount of data that can be accumulated from each and every element of a wind power system in a real-time environment.

References

1. Kalaiarasi, D., Anusha, A., Berslin Jeni, D., & Monisha, M.. (2016, April). Enhancement of hybrid power systems using IoT. *International Journal of Advanced Research Trends in Engineering and Technology (IJARTET)*, 3(19).
2. Buyya, R., Calheiros, R. N., & Dastjerdi, A. V. (Eds.). (2016). *Big data: principles and paradigms*. Cambridge: Morgan Kaufmann.
3. Byun, J., Kim, S., Sa, J., Kim, S., Shin, Y.-T., & Kim, J.-B. (2016). Smart city implementation models based on IoT technology. *Advanced Science and Technology Letters, 129*(41), 209–212.
4. Cao, J., & Yang, M. (2013). Energy internet—towards smart grid 2.0. In *2013 Fourth international conference on networking and distributed computing* (pp. 105–110). IEEE.
5. Ritzer, G. (2015). Automating prosumption: The decline of the prosumer and the rise of the prosuming machines. *Journal of Consumer Culture, 15*(3), 407–424.
6. Zhang, Y., Xiang, Y., & Wang, L. (2016). Power system reliability assessment incorporating cyber-attacks against wind farm energy management systems. *IEEE Transactions on Smart Grid, 8*(5), 2343–2357.
7. Hatzivasilis, G., Fysarakis, K., Soultatos, O., Askoxylakis, I., Papaefstathiou, I., & Demetriou, G. (2018). The industrial internet of things as an enabler for a circular economy Hy-LP: A novel IIoT protocol, evaluated on a wind park's SDN/NFV-enabled 5G industrial network. *Computer Communications, 119*, 127–137.
8. Lotfi, H., & Khodaei, A. (2016). Levelized cost of energy calculations for microgrids. In *2016 IEEE Power and Energy Society General Meeting (PESGM)* (pp. 1–5). IEEE.
9. Alhmoud, L., & Al-Zoubi, H. IoTapplications in wind energy conversion systems. *Open Engineering, 9*(1), 490–499.
10. Moness, M., & Moustafa, A. M. (2015). A survey of cyber-physical advances and challenges of wind energy conversion systems: prospects for internet of energy. *IEEE Internet of Things Journal, 3*(2), 134–145.

Chapter 15
An Application of IoT to Develop Concept of Smart Remote Monitoring System

Meera Sharma, Manish Kumar Singla, Parag Nijhawan, Souvik Ganguli (i),
and S. Suman Rajest

15.1 Introduction

One of the main contributors of clean energy all over the world, currently, is the solar photovoltaic system. A PV system's power generation potential is the main factor to be determined [1]. This potential may vary depending upon various constraints such as the technology employed as well as the location considered. The expense of every kWh pay backs the return on investment on the electricity potential [2]. Hence, before the installation of PV systems, certain measures need to be taken, so as to have larger energy potentials. There may be chances of collapse and maintenance issues at the time of operation of the system despite the efforts made during or before the installation of the PV systems. Such problems are occurring more often where the installations are in remote locations. Therefore, to mitigate such problems, a suitable approach is desired. The most accurate way to avoid such issues is the frequent check that is almost impossible for a person to carry out. Even problems involving concentration, giving full attention, and identification of solution cannot be encountered [3].

The best way to cater such a problem is to adopt the most emergent method known as the "Internet of Things" (IoT) for remote evaluation of models. IoT helps us to interact with objects used in our daily life in a much quicker and easy way with the usage of communication devices following network protocols [4, 5].

IoT has a vast range of applications including infrastructure, industrial automation, healthcare support, home power management and the renewable energy framework, traffic maintenance, automotive enterprise, micro-grids, and intelligent drive

M. Sharma · M. K. Singla · P. Nijhawan · S. Ganguli (✉)
Thapar Institute of Engineering and Technology, Patiala, India
e-mail: souvik.ganguli@thapar.edu

S. S. Rajest
Vels Institute of Science, Technology & Advanced Studies, Chennai, India

© Springer Nature Switzerland AG 2020   233
A. Haldorai et al. (eds.), *Business Intelligence for Enterprise Internet of Things*,
EAI/Springer Innovations in Communication and Computing,
https://doi.org/10.1007/978-3-030-44407-5_15

systems, among others [5, 6]. Based on the advent of the IoT, solar PV systems are the latest targets being focused upon. This is because of their increasing usage in the current energy sector trends especially in energy distribution. Solar PV system's popularity would be a major breakthrough for the implementation of IoT systems in combination with it, thereby giving an edge for the IoT service suppliers as well as the consumers.

15.2 Photovoltaic Systems

A PV system consists of the arrangement of a PV module, power converters, and storage devices. In essence it is the power harvester, which changes sunlight into electricity [7]. This technique is quite different from the traditional process which involves fossil fuels for power generation. Although power transmission and distribution embraces similar traditional methods, PV arrays are formed by grouping PV modules; such PV arrays are known as PV generators when arranged in series and parallel configurations [8–10]. They are then installed in a manner such that they are exposed to direct sunlight. The DC electricity generated from the PV generator is changed into AC with the assistance of the inverters. This power can be self-consumed or distributed to the energy grid through the transmission network [11]. Nonetheless, the energy can be stored using the batteries instead of being transferred. The PV models are grouped into two forms, i.e., the off-grid and the on-grid PV models, centered on various forms of functional components. An illustration in Fig. 15.1 provides a layout of the operation of the photovoltaic system.

Fig. 15.1 Layout of a typical photovoltaic system [12]

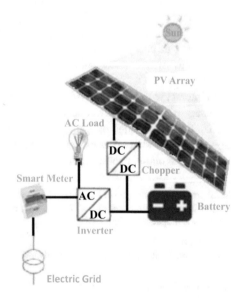

15.2.1   IoT and Its Requirement in Photovoltaic Systems

Internet of Things, abbreviated as IoT, is a technology that is developed by grouping together "wireless technologies, micro-electromechanical systems, and the Internet" [3–6]. The mechanical/digital machines, computing devices and objects, solitary identifiers, and other such analogous things coordinating together constitute IoT. Because of this synchronization, which transmits the data across the network, the distance between the operational systems and information technology is closed without the aid of human-to-human and human-to-machine interaction.

The contemporary science and engineering systems cannot solve the most intricate issues which IoT can solve. The operational behavior of various components of the PV systems, which are used for generating power, varies. In short, a constant generation of power is not achievable throughout because of the solar intensity being weather dependent and time varying [12]. This has an indirect effect on the working of other components of the device such as voltage levels of power converters, status of battery charging, and load energy demands. Some environmental conditions, such as accumulation of dust, are also sometimes responsible for the poor performance of the PV system. Nevertheless, these problems can lead to the collapse of the whole system in longer terms. In order to maintain the operating data log, humans face difficulties to monitoring since it requires visits to the plant site time and time again. Henceforth, humans consume a lot of time in addressing these failures because of system breakdown or bad performance [3]. To check the parameters of the system and store them in cloud, a continuous monitoring system together with the PV system is to be equipped. This stored data will provide the performance parameters and the causes of poor performance and will make troubleshooting and maintenance operation much easier and faster. Therefore, the need for IoT is necessary to optimize the device parameters with the option of remote control.

15.2.2   IoT-Based Photovoltaic System Architecture

The IoT architecture of a photovoltaic system consists of three distinct layers. Figure 15.2 clearly displays the IoT photovoltaic system architecture. The initial layer incorporates the PV model design ambience and is connected for user satisfaction according to the required configurations. In this particular case, the Arduino server is interlinked with the components of the PV device, creating the second layer of the IoT architecture. Along with an Internet firewall option, using a router, the web server can be inter-connected with the hardware projects of the PV scheme, hence forming the gateway linkage in this second layer. The Arduino server is majorly responsible for this integration. It carries out the main functions of controlling, monitoring, and managing the PV scheme hardware constituents. The server collects the data from the third and last layer known as the remote monitoring and control layer. This information is transmitted to the storage devices that

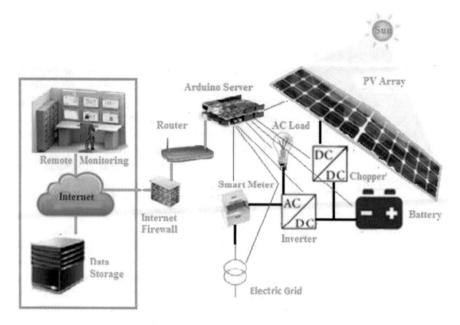

Fig. 15.2 IoT-aided layout for photovoltaic system [12]

help generate periodic reports. Using an Android interface with cloud data storage via a Wi-Fi network, these data can be drawn up in the form of visual graphs and reports, and then the users can access it accordingly.

15.2.3 Proposed Concept for IoT-Aided System for Photovoltaic Monitoring

This research proposes a device that evaluates the condition of the PV framework based on the IoT-centered network with the purpose of remotely controlling it. For the transmission of sensor information, a mobile radio network is used. The remote server data is sent through a GPRS module [13]. Figure 15.3 displays IoT technology schematics for a solar power plant.

A three-layered schematic diagram having the bottom layer as the sensing layer consists of current and voltage sensors, irradiance-measuring device (pyranometer), and other sensors. The sensing layer also consists of a microcontroller-based data processing which is acquired by the sensors. A wireless module is utilized to communicate with the microcontroller in order to initialize and start transmitting data to the server.

The second layer, known as the system layer, is where information logging is done from the plant for real-time transmission and includes database storage as

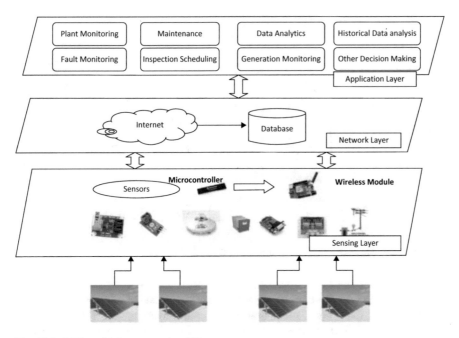

Fig. 15.3 IoT-based solar power plant [7]

well [14]. The application layer further uses this stored and processed data from the network layer. Based on the collected data's processing and storage, the web-based services are hence designed smartly. In order to help in monitoring the plant's performance, graphical user interfaces are employed. With the console, the decision-making time is shortened, indicating the administrator with historical data-based decision.

A remote monitoring system based on IoT makes it much easier to supervise the solar power plant's performance as a whole using a web-based technique as shown in Fig. 15.4.

15.3   Summary

The main advantage of using IoT photovoltaic technology is that we can accurately view the status of our property from the central control panel. Through connecting the computer to the cloud network, we can even pinpoint where the problem lies and allow technicians to repair it long before the entire system breaks down. The network is less vulnerable to the production issues (due to power outages) and potential security threats through the use of the Internet of Things. Through install-ing an IoT solution directly and linking solar devices, we can monitor our solar

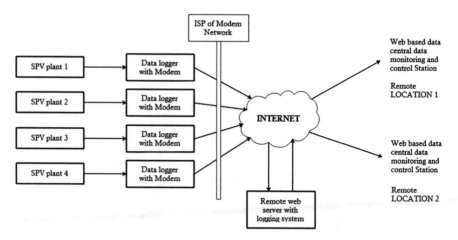

Fig. 15.4 Complete schematic of the smart remote monitoring system [7]

power system's broadband even when there are thousands of devices connected to the network. In addition to real-time business notifications, the solar industry's Internet of Things increases energy efficiency and profitability by gathering historical modeling data. This makes energy generation more efficient in terms of both costs and logistics.

References

1. Cota, O. D., & Kumar, N. M. (2015). Solar energy: a solution for street lighting and water pumping in rural areas of Nigeria. *Proceedings of International Conference on Modelling, Simulation and Control (ICMSC-2015), 2*, 1073–1077.
2. Pezzotta, G., Pinto, R., Pirola, F., & Ouretani, M. Z. (2014). Balancing product-service provider's performance and customer's value: the service engineering methodology (SEEM). In *6th CIRP conference on industrial product-service systems* (Vol. 16, pp. 50–55). Elsevier.
3. Internet of Things (IoT). *IoT Agenda.* http://internetofthingsagenda.techtarget.com/definition/Internet-ofThings-IoT.
4. Atzori, L., Iera, A., & Morabito, G. (2010). The internet of things: A survey. *Computer Networks, 54*(15), 2787–2805.
5. Zanella, A., Bui, N., Castellani, A., Vangelista, L., & Zorzi, M. (2014). Internet of things for smart cities. *IEEE Internet of Things Journal, 1*(1), 22–32.
6. Bellavista, P., Cardone, G., Corradi, A., & Foschini, L. (2013). Convergence of MANET and WSN in IoT urban scenarios. *IEEE Sensors Journal, 13*(10), 3558–3567.
7. Adhya, S., Saha, D., Das, A., Jana, J., & Saha, H. (2016). An IoT based smart solar photovoltaic remote monitoring and control unit. In *2016 2nd international conference on Control, Instrumentation, Energy & Communication (CIEC)* (pp. 432–436). IEEE.
8. Reddy, B. V., Sata, S. P., Reddy, S. K., & Babu, B. J. (2016). Solar energy analytics using internet of things. *International Journal of Applied Engineering Research, 11*(7), 4803–4806.

 9. Nadpurohit, B., Kulkarni, R., Matager, K., Devar, N., Karnawadi, R., & Carvalho, E. (2017). Iot enabled smart solar pv system. *International Journal of Innovative Research in Computer and Communication Engineering, 5*(6), 11324–11328.
10. Guamán, J., Guevara, D., Vargas, C., Ríos, A., & Nogales, R. (2017). Solar manager: Acquisition, treatment and isolated photovoltaic system information visualization cloud platform. *International Journal of Renewable Energy Research, 7*(1), 214–223.
11. Vignesh, R., & Samydurai, A. (2017). Automatic monitoring and lifetime detection of solar panels using internet of things. *International Journal of Innovative Research in Computer and Communication Engineering, 5*(4), 7014–7020.
12. Kim, J., Byun, J., Jeong, D., Choi, M.-i., Kang, B., & Park, S. (2015). An IoT-based home energy management system over dynamic home area networks. *International Journal of Distributed Sensor Networks, 11*(10), 828023.
13. Chen, X., Sun, L., Zhu, H., Zhen, Y., & Chen, H. (2012). Application of internet of things in power-line monitoring. In *2012 International conference on cyber-enabled distributed computing and knowledge discovery* (pp. 423–426). IEEE.
14. Peijiang, C., & Xuehua, J. (2008). Design and Implementation of Remote monitoring system based on GSM. In *2008 IEEE Pacific-Asia workshop on computational intelligence and industrial application* (Vol. 1, pp. 678–681). IEEE.

Chapter 16
Heat Maps for Human Group Activity in Academic Blocks

Rajkumar Rajasekaran, Fiza Rasool, Sparsh Srivastava, Jolly Masih, and S. Suman Rajest

16.1 Introduction

Sensing and identifying human group activities has garnered a rising research interest in many practical applications like video surveillance and crime detection. Multifarious algorithms have been designed and proposed for the recognition of group activities or interactions. Though many of the already prevailing algorithms have unearthed the wide-ranging features among the activities such as the full movement data, they were unsuccessful in obtaining information about the temporal motion (e.g., they failed to depict the time and place of a person) [1]. Hence, they suffered from restrictions in case of detecting and identifying more convoluted group activities. Though these other techniques include and study the temporal data by using the chain models like HMM (hidden Markov models), they still suffer from one disadvantage which is the requirement of enormous amounts of training data, along with managing the variations in motion operation and behavior. Also, several uncertainties also play a challenging concern in recognition of group activity. Due to inherent variation in the motion of people in cluster activities, the precision and correctness of recognition might be prominently influenced by the unsure changing characteristic of group motion [2].

The ultimate goal of the system is optimization of targeted digital advertising. It is the most effective form of advertising [3]. Most advertising done these days is online, and almost all of it is targeted. Targeted advertising is a method by which

R. Rajasekaran (✉) · F. Rasool · S. Srivastava
Vellore Institute of Technology, Vellore, India

J. Masih
Erasmus School of Economics, Rotterdam, Netherlands

S. S. Rajest
Vels Institute of Science, Technology & Advanced Studies, Chennai, India

© Springer Nature Switzerland AG 2020
A. Haldorai et al. (eds.), *Business Intelligence for Enterprise Internet of Things*,
EAI/Springer Innovations in Communication and Computing,
https://doi.org/10.1007/978-3-030-44407-5_16

advertisers are able to provide more specific and relevant advertisements to users based on data they have collected on the users from various sources – services like Google AdWords and Facebook advertising. The reason why this type of advertising is extremely effective is because it utilizes data that is most pertinent to the user or taker. Through this chapter, we put forth a very innovative heat map algorithm for the recognition of human group activity in academic buildings and hostel blocks. The main objectives of this chapter can be stated as the following points:

1. This chapter presents a completely new heat map (HM) aspect to characterize group activities in academic buildings and hostel blocks. This proposed HM will be able to efficiently fetch the temporal motion data of the crowd activities based on the temporal and cultural factors [4].
2. This chapter proposes to implement a thermal diffusion method to generate the heat map. In this technique, we will be able to efficiently address the motion uncertainty from different people.
3. We also intend to perform a new SF (surface fitting) method which can also identify the group activities. This presented process can essentially study the features of the heat map and can provide us with recognition results effectively.

16.2   Literature Survey

Anandakumar and Umamaheswari [5], Arizona State University, USA, applied a dynamic, actual-time heat map data capture, and for two military simulations, they generated tools which include unmanned aerial vehicle sensor operator scenario and ground-based combat scenario. Arulmurugan and Anandakumar [6] have developed SURV or Smart Urban Visualization system. SURV is a new framework developed by applying HTML, CSS, as well as Mapbox.js and CARTO.js JavaScript libraries in order to generate dynamic visualizations and interactive maps. Python, the most popular versatile language of these times, is also integrated and used with this system due to its excellent performance in data processing. SURV offers an interface or a web app where the end-user will be able to perform three-dimensional (3D) visualizations along with data animation to deliver better data analysis.

Hirenkumar Gami, Miami University, USA, has only used a single PIR sensor combined with machine learning algorithms and signal processing to clearly study and predict the presence, proximity, and direction of the crowd movements on the floors of a building. [7], PSUT, Jordan, presented the significance of visualization techniques when implemented on huge real-time data sets from a certain corporation in Jordan as detailed tools for analysis of data. The name of the software utilized to analyze and visualize this data is known as Tableau. [8] in his research paper perceives the pairwise actions by fetching the causality features from the bitrajectories. Ni et al. further extended the cause and effect characteristics into three different kinds. Among these, the first is individuals, the second is pairs, and the third is groups. [9] discovers and studies crowd activities by presenting connected operating segmentations in order to depict and display the corresponding congruence among the people. Cheng et al. propose to study the pattern for group activity by taking into consideration the Gaussian parameters and creating trajectories that are determined by the calculation from many people.

16.2.1 Detailed Working

The system is composed of two components – the hardware detectors and the software backend. The hardware components have been prototyped using an Arduino development board with PIR sensors. These are fixed at certain key positions in the academic buildings, and based on the factors we considered, a number of people visiting the academic buildings are stored in a software database. And using the surveillance cameras in the academic buildings, the activities of students and staff are continuously recorded. As discussed before, if the overall characteristics are straight away being obtained from the motion data, most of the temporal data/information will be lost. To avoid this loss of information, it is proposed that the trajectory is represented as a sequence of heat sources. So, for converting the trajectory to the series of heat sources, we need to firstly dissect the full video scene into tiny patches which are non-overlaying. In the case wherein a trajectory goes falling into a patch, this case will be regarded as a heat source. In this way, we can convert a trajectory into several heat source series, and the thermal values pertaining to the series of heat sources can be organized in an increasing order following the course of the trajectory [10]. The obtained temporal data converted into some useful information is effectively inserted.

Besides, as the trajectories of people might show huge deviations, straight away utilizing the series of heat sources by way of characteristics shall be significantly influenced by the motion fluctuation. Therefore, to reduce the fluctuation in motion, a consecutive technique can be implemented by introducing a further process in the form of thermal diffusion which disperses the heat from the series of heat sources. This is called diffusion result of the heat map. Although this process has been used in the past, this is the first time we are using it for human activity recognition in academic buildings and hostel blocks. With our proposed heat map feature, we can describe the activities' information of motion with the help of 3D surfaces. But again, the problem occurs in choosing an appropriate technique for implementing recognition which uses this feature of HM. Therefore, we additionally proposed the SF (surface fitting) method for recognition of activities [11]. In the proposed SF technique, firstly, a bunch of prevailing classes of surfaces are recognized in order to represent the diverse activities. Secondly, the various parallels concerning the input surface HM and the standard surface classes can be easily determined. In the final step, the standard surface class that almost matches will be selected, and its equivalent human activity will be declared as the activity which is most recognized for the inputted heat map. Therefore, the data that is generated from HMs is passed onto the digital advertising partners who then utilize this information to serve the targeted advertising.

16.2.2 Hmb Method

The working of the heat map-based algorithm is shown below. Firstly, the input trajectories are converted into a series of heat sources, followed by the process of thermal diffusion to generate the HM characteristic which calculates the group activity input [12]. After this, the surface fitting technique is useful for implementing the activity recognition. The thermal diffusion and the transfer of heat source series, along with the SF method, are the prime components of the presented algorithm shown in Fig. 16.1.

16.3 Experimental Results

The experimental outcome for the proposed heat map-based algorithm is discussed here. In the experiment, the HMB algorithm is compared with three other methods: the WF-SVM algorithm, the GRAD algorithm, and the PGTB algorithm. Here, we performed four individual experiments and the average of the results is calculated. Further, we have determined miss and false alarm (FA) for the four different methods and then calculated their total as shown in Table 16.1.

From the above table, it is clear that the PGTB algorithm fails to yield good results. Comparatively, the WF-SVM and the GRAD algorithms provide better results and performance with the help of many features which are distinguishable. When a comparison is made among these mentioned methods, the proposed HMB algorithm shows the finest performance. And, as activities are detected, we can now send the data to the advertising partner who can advertise based on the people's interests and activities (Fig. 16.2).

16.3.1 Analysis of Human Footfall in an Academic Building (in Terms of Density)

From the above table, we can see that the density of human footfall is high toward the morning of each day. This density in the early morning stays maximum for the entire day except in the case of Monday, where 10 am–2 pm has increased density over the morning 6 am–10am slot. For almost all the days, human density dwindles as the day progresses with least amount of density at nighttime. One more noticeable change takes place on Wednesday, 10 am–2 pm slot, where the density falls

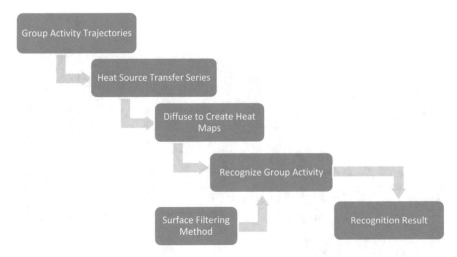

Fig. 16.1 Heat map method

Table 16.1 Heat map result

	HMB	WF-SVM	PGTB	GRAD
Miss	0.9%	11.8%	33.1%	10.7%
FA	1.2%	0.4%	4.2%	1.4%
Miss	5.6%	8.5%	8.1%	16.5%
FA	0.8%	2.1%	34.5%	0.9%
Miss	4.7%	8.8%	31.9%	13.9%
FA	0.8%	1.9%	0.1%	2.1%
Miss	3.2%	5.2%	34.8%	8.5%
FA	0.3%	0.3%	6.8%	0.7%
Miss	2.9%	4.7%	45.1%	9.9%
FA	1.3%	1.7%	0.02%	2.8%
Miss	5.2%	3.1%	59.8%	2.7%
FA	0.5%	2.0%	0.3%	1.1%
	3.8%	6.9%	36.9%	11.2%

sharply as no classes are scheduled during this reserved slot meant for non-academic activities as shown in Table 16.2.

The following can be visualized better using the following heat map:

Thus, this heat map shown in Fig. 16.3 helps us to visualize that the density of people in the academic building is maximum during the working hours which is indicated by the bright red shades. As the day progresses, the pattern of the heat map turns toward lighter pastel colors when almost all the students have returned to their hostels and the footfall is negligible.

Fig. 16.2 Human footfall in an academic block

Table 16.2 Day calculations 1

Days	6 am–10 am	10 am–2 pm	2 pm–6 pm	6 pm–10 pm	10 pm–6 am
Monday	0.64	0.7	0.51	0.32	0.0012
Tuesday	0.73	0.57	0.42	0.21	0.001
Wednesday	0.82	0.35	0.56	0.41	0.0015
Thursday	0.58	0.64	0.55	0.35	0.001
Friday	0.61	0.52	0.61	0.18	0.003

The bar graphs and the line graphs presented in Fig. 16.4 provide a better picture in visualizing the footfall of the students in the building. These can prove to be really helpful to target the temporal aspects of the situation and to tackle any mishap as shown in Fig. 16.5 Table 16.3.

16.3.2 Visualization for Hostel Blocks

Here, we observe that the data collected through the sensors is at its maximum during the nighttime between 10 pm and 6 am when nearly every student has returned to the hostel post in time. The density from 6 am to 10 pm is around 0.5 as the

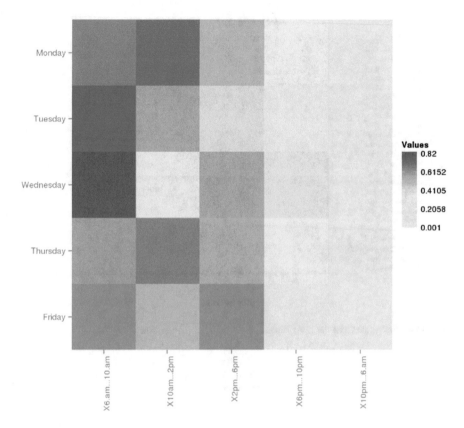

Fig. 16.3   Heat map

classes start from 8 am. The 10 am–2 pm slot also sees a rise in density as most of the students return to the hostel during lunch hours. Overall, the heat map for human footfall at hostels can be plotted as follows in Fig. 16.6:

Here, a line graph shown in Fig. 16.7 and the bar plots provide a better insight to the data collected by the PIR sensors. The overall human footfall through these visualizations can prove to be extremely beneficial in the process of group activity detection, analysis, and recognition as shown in Fig. 16.8.

16.4   Summary

Thus, through this chapter, we have tried to devise the HMB algorithm, which is an algorithm based on heat maps, for the purpose of detecting and identifying group activities, in order to avoid loss of life and property in case of a mishap. The use of PIR sensors which are exceptional devices for detecting presence of humans, with the help of a minor form-factor and strong-featured design, is the best choice for

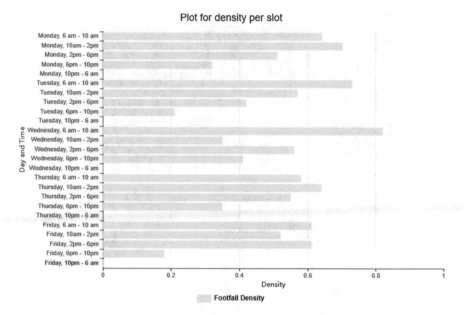

Fig. 16.4 Bar graphs and the line graphs

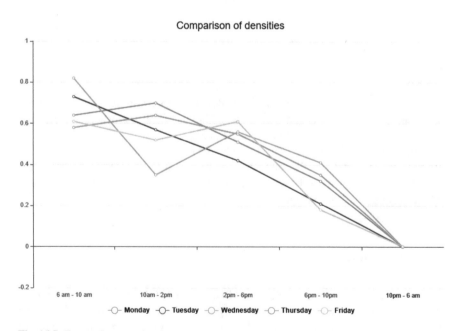

Fig. 16.5 Foot path comparison

Table 16.3 Day calculations 2

Days	6 am–10 am	10 am–2 pm	2 pm–6 pm	6 pm–10 pm	10 pm–6 am
Monday	0.49	0.38	0.33	0.69	0.9988
Tuesday	0.51	0.41	0.38	0.78	0.999
Wednesday	0.52	0.68	0.36	0.66	0.9985
Thursday	0.48	0.54	0.32	0.68	0.999
Friday	0.46	0.56	0.37	0.82	0.997

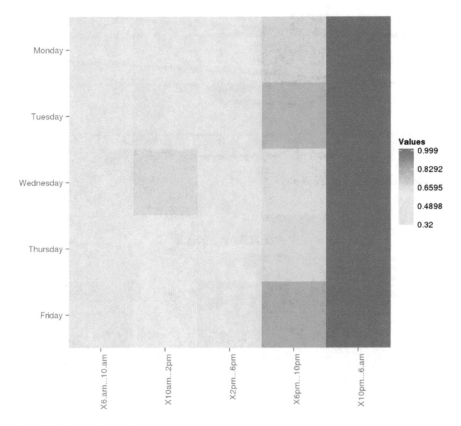

Fig. 16.6 Human foot path

achieving cost-efficient surveillance. We, therefore, propose to generate the heat map, for the purpose of demonstrating the group activities, and then consequently implement the SF method for recognition of crowd activity. Based on the results, we can effectively avoid any mishap based on the density and activity detection of the people and hence can save many precious lives.

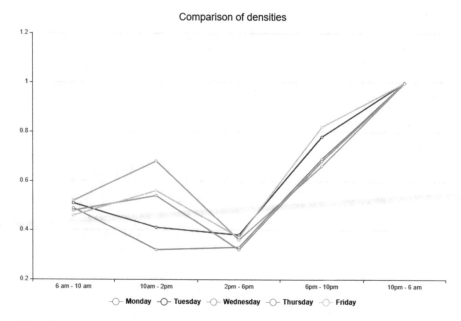

Fig. 16.7 Human foot path densities

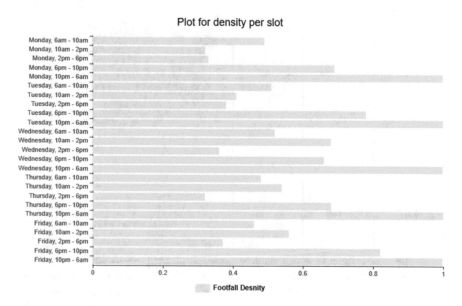

Fig.16.8 Overall analysis

References

1. Chu, H., Lin, W., Wu, J., Zhou, X., Chen, Y., & Li, H. A new heat-map-based algorithm for human group activity recognition. *ACM*.
2. Amresh, A., Femiani, J., Fairfield, J., & Fairfield, A. (2013, July). UAV sensor operator training enhancement through heat map analysis. In *Information Visualisation (IV), 2013 17th international conference*. IEEE.
3. Ihaddadene, N., & Djeraba, C. (2008). Real-time crowd motion analysis. In *2008. ICPR 2008. 19th international conference on pattern recognition* (pp. 1–4). IEEE.
4. Anandakumar, H., & Umamaheswari, K. (Mar. 2017). Supervised machine learning techniques in cognitive radio networks during cooperative spectrum handovers. *Cluster Computing, 20*(2), 1505–1515.
5. Anandakumar, H., & Umamaheswari, K. (Oct. 2018). A bio-inspired swarm intelligence technique for social aware cognitive radio handovers. *Computers & Electrical Engineering, 71*, 925–937. https://doi.org/10.1016/j.compeleceng.2017.09.016.
6. Arulmurugan, R., & Anandakumar, H. (2018). Early detection of lung Cancer using wavelet feature descriptor and feed forward Back propagation neural networks classifier. *Lecture Notes in Computational Vision and Biomechanics*, 103–110. https://doi.org/10.1007/978-3-319-71767-8_9.
7. Haldorai, A. R., & Murugan, S. Social aware cognitive radio networks. In *Social network analytics for contemporary business organizations* (pp. 188–202). https://doi.org/10.4018/978-1-5225-5097-6.ch010.
8. Arulmurugan, R., & Anandakumar, H. (2018). Region-based seed point cell segmentation and detection for biomedical image analysis. *International Journal of Biomedical Engineering and Technology, 27*(4), 273.
9. Suganya, M., & Anandakumar, H.. (2013, December). Handover based spectrum allocation in cognitive radio networks. In *2013 International Conference on Green Computing, Communication and Conservation of Energy (ICGCE)*. https://doi.org/10.1109/icgce.2013.6823431. https://doi.org/10.4018/978-1-5225-5246-8.ch012
10. Huacón, C. F., & Pelegrin, L. (2018). SURV: A system for massive urban data visualization. In *URTC 2017*. IEEE MIT.
11. Gami, H. (2017). Movement direction and distance classification using asingle PIR sensor. In *IEEE sensors letters*. IEEE.
12. Sharawi, L. I., & Sammour, G. Utilization of data visualization for knowledge discovery in modern logistic service companies. In *SENSET 2017*. IEEE.

Chapter 17
Emphasizing on Space Complexity in Enterprise Social Networks for the Investigation of Link Prediction Using Hybrid Approach

J. Gowri Thangam and A. Sankar

17.1 Introduction

Social Network is a structure comprised of a lot of actors. It provides a set of methods for analyzing the structure of whole social entities as well as a variety of theories explaining the patterns are pragmatic in these structures. By using those structures, local and global patterns and visible entities can be identified and the network dynamics can be examined.

Social Network Analysis (SNA) [1] is based on the assumption of the importance of relationships among interacting units. In recent decades, online social networks [2] have turned out to be progressively significant assets for individual interaction, information process and social impact dispersion. Understanding and modeling the mechanisms by which these networks evolve becomes a fundamental issue and dynamic research area. Therefore, the conviction of SNA has pulled in significant intrigue and interest from social network community. Quite a bit of this intrigue can be credited to the engaging focal point of social network investigation on connections among social elements and ramifications of these connections.

It has become very significant in today's scenario because of the accomplishment of online person to person communication. This makes use of network theory [3] to analyze social networks. SNA views social relationships in terms of network theory, consisting of nodes, representing individual actors within the network, and ties or edges that represent relationships between the individuals, such as friendship, kinship, and organizations. These networks are often depicted in a social network diagram or sociograph, where nodes are represented as points and ties are represented as lines.

J. G. Thangam (✉) · A. Sankar
Department of Computer Applications, PSG College of Technology,
Coimbatore, Tamil Nadu, India

© Springer Nature Switzerland AG 2020

253

A. Haldorai et al. (eds.), *Business Intelligence for Enterprise Internet of Things*,
EAI/Springer Innovations in Communication and Computing,
https://doi.org/10.1007/978-3-030-44407-5_17

Link Prediction [4] is a noteworthy issue in social network. It is the problem of predicting the existence of link between two entities. Since the social network is vast, it is very difficult to predict the link at the earliest. So there is real urge to consider about the efficiency of the algorithm in terms of time and space. This technique is to reduce the dimensions so that link can be rapidly predicted.

This chapter is organized as follows: the related works is presented in next section, while the preliminaries, the problem description, and proposed work are presented in Sects. 17.3, 17.4, and 17.5. In Sect. 17.6, an experimental study is conducted on real data sets. The social network centrality measures are described in Sect. 17.7. The complexity analysis is represented in Sect. 17.8. The evaluation metrics are analyzed in Sect. 17.9. Finally, Sect. 17.10 is conclusion and the future work.

17.2 Related Works

A Hybrid Active Learning Link Prediction (HALLP) [5] has used machine learning techniques for binary classification of links. D. Sharma [6] has made a test examination of the connection forecast procedures. Michael Fire et al. [7] have used topological feature (friend's measure) for predicting the links. A novel strategy has been proposed by Naveen Gupta et al. [8], for identifying the missing links by considering the uncommon neighbors. Manu Kurakar et al. [9] have identified complex diseases based on the sequence similarity in protein networks.

A link prediction model has been proposed by Wal et al. [10], for identifying the links that might appear and disappear in future. A friend recommendation has been suggested by traversing all paths of limited length by Alexis et al. [11]. A Social Attribute Network (SAN) approach proposed by Neil Zhengiang Gong et al. [12], have considered the performance of supervised and unsupervised algorithms. A novel approach proposed by Aditya Krishna Menon et al. [13] has addressed the class imbalance problem by optimizing the ranking loss. Daniel et al. [14] have proposed a tensor-based method for predicting the future links temporally.

Jichang Zhao et al. [15] evaluated the performance of link prediction based on local information by using sampling techniques. Ryan N. Lichtenwalter [16] proposed a machine learning algorithm with class imbalance. An algorithm for parameter reduction of soft binary relation was proposed and investigated by Guangh Yu [17]. Xiuqin Ma [18] proposed a parameter reduction algorithm of soft set for decision making.

17.3 Preliminaries

The notion of a social network and the methods of SNA have attracted considerable interest and curiosity from the social and behavioral science community in recent decades. It is used to quantify the relationships among social entities and on the patterns and implications of these relationships. Milgram's small-world phenomenon is revealed in social network. It states that the average path length between any two nodes should be shorter. The social network is represented as a graph.

17.3.1   Graph

A graph G is defined as $G = (V, E)$, where V is a set of vertices and E is a set of edges. The notion of vertices and nodes, links and edges is used interchangeably in the literature. The adjacent matrix is used for representing any graph. The adjacency matrix of a graph G with n vertices is an $n * n$ binary matrix given by $A = [a_{ij}]$ defined as $a_{ij} = 1$ if the ith and j^{th} vertices are adjacent, that is, there is an edge connecting the ith and j^{th} vertices, and $a_{ij} = 0$, otherwise, that is, if there is no edge connecting the ith and j^{th} vertices.

The graph shown in Fig. 17.1 is a simple undirected graph with set of vertices as $\{A, B, C, D, E, F\}$ and set of edges as $\{AB, AC, BF, BD, BC, CE\}$. An edge from a node A to node B is called a path. It is of length 1 and known to be length-1 path. In a similar manner, the path from A node D through node B is of length 2 which is referred as length-2 path.

17.3.2   Link Prediction Problem

Link Prediction Problem [19, 20] is identifying a potential or possible link among vertices or nodes in a network. It is defined as given a snapshot of a social network, where new interactions among its members that are likely to occur in the near future can be inferred. Predicting such links is not an easy task. Many similarity index-based link prediction methods have been based on nature of computing them, as local structure, global structure, and quasi local structure. Friendlink algorithm is based on Quasi-local structure measure. It uses path of length more than 2.

Fig. 17.1 Graph

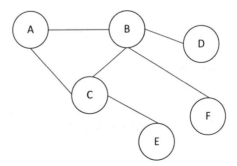

17.4   Problem Description

The link prediction is an issue for foreseeing whether the link will exist in future or not. The similarity score between the nodes is calculated for predicting the social links. Since the social network data set is enormous, there is a need for dimensionality reduction, without affecting the consistency of the data set. Soft set theory is used for dimensionality reduction. The reduced dimension is used for predicting the future links. The similarity score between the nodes are calculated for the reduced dimension. Based on this similarity score, ranking of all other nodes has to be done with respect to a particular node. The future link is more probable to occur, if the similarity score between the nodes is higher.

17.4.1   Soft Set

Molodtsov [21] soft set theory is an overview of fuzzy set theory [22] that deals with uncertainty in a nonparametric manner. Let U be an initial universe and E be the set of possible parameters. The power set of U is denoted by $P(U)$. The Soft set is defined as follows:

Definition A pair (F,E) is called a soft set over U, where F is a mapping of E into $P(U)$, that is,

$$F, E \rightarrow P(U) \tag{17.1}$$

In a soft set theory, the reduction of parameter is a vital problem. It deals with the elimination of frequent occurring parameters in order to take an optimal decision making, without affecting the novelty of the data set, so that the number of parameters is reduced to simply the calculations. Hence the soft set is used to discover the similar links between the nodes. These similar links are removed from data set that yields in reducing the dimension. Thereby the space as well time complexities will have a significant improvement.

17.4.2   Friendlink Algorithm

Friendlink algorithm [23] finds the similarities between nodes in a sociograph using the adjacency matrix. Rubin's algorithm [24] is used to find the paths of length l.

Definition The similarity score $sim(u_x, u_y)$ between two nodes u_x and u_y is defined as counting of paths of varying length l from u_x to u_y is as given below:

$$\text{sim}\left(u_x, u_y\right) = \sum_{k=2}^{l} \frac{1}{k-1} * \frac{\left|\text{paths}_{u_x, u_y}^k\right|}{\prod_{m=2}^{k}\left(n-m\right)} \qquad (17.2)$$

where n is the number of vertices in a graph, l is the maximum length of a path from u_x and u_y. $\frac{1}{k-1}$ is an attenuation factor that weighs paths according to their length. $\left|\text{paths}_{u_x, u_y}^k\right|$ is the number of all length-l paths from u_x to u_y. $\prod_{m=2}^{k}\left(n-m\right)$ is the number of all possible length-l paths from u_x to u_y. k and m denote current iteration. For example, if the two nodes are similar, the similarity score is expected to be close to 1. On the other hand, if the two nodes are not similar, the similarity score is expected to be close to 0.

The Friendlink algorithm comprises of two functionalities, one is for computing the path of length-2 between two nonexistent nodes and other is for computing the similarities between them. Then the similarity scores will be ranked. Finally the node with the highest similarity score will be declared as the node which is more probable to be communicated in future. Hence nodes can be recommended to the target node according to the similarity scores.

17.5 Proposed Work

Link prediction is a pivotal problem in SNA in order to be acquainted with associations between nodes in any social communities. The proposed approach is explained in Fig. 17.2.

The Social Network data set is obtained as edge list for a particular network. The edge list is then converted into an adjacency representation. The soft set is applied and the preprocessed data is obtained. The Soft set and Friendlink-based Link Prediction algorithm (SFLP) uses the Friendlink algorithm for the path length of 2 and 3. The similarity score is calculated for the original data set using the length-2 path and length-3 path. Ultimately, the similarity score of a node against all other nodes are computed and similarity scores are ranked. The node with higher similarity score is the node to occur in future which is predicted. The pseudo code for ranking the nodes is provided in Algorithm 1.

Algorithm 1: Soft Set and Friendlink-Based Link Prediction (SFLP) Algorithm
Input: Adjacency Matrix.
Output: Ranked prediction list

1. Reduce the dimension using soft set.

   For each node $node_i$.

   Find all the links such that $a_{ij} > 0$
   Remove a_{ij} if the link is redundant
   End For

2. Compute similarity score $sim(node_i, node_j)$ using Friendlink algorithm.

   For each node $node_i$

   If $a_{ij}! = 1$
   Find the similarity scores with all   other nodes.
   End if
   End For

3. Rank the nodes.

The algorithm is first organized to reduce the dimensions of the dataset. The reduced data set is used for finding the similarity score between the nodes. Then all the similarity scores are ranked, and the node with the maximum similarity score value will be the node to be linked in future. The similarity scores obtained using SFLP is compared with the similarity score of **F**riendlink **L**ink **P**rediction (FLP) algorithm. Those results are analyzed against the social network centrality measures.

17.6   Experimental Setup and Results

The experiments are conducted on a 2.50-GHz Intel Dual core PC with 4GB RAM running Microsoft 7 ultimate. The SFLP and FLP algorithms are implemented using MATLAB. An examination is made to assess the exhibition of the proposed method.

   To evaluate the efficiency of the algorithm, it is examined using real world data set. The data set is loaded from UCI Network Data Repository. The Zachary's Karate Club data set is a social network of friendships of a karate club at a US university. Each node represents a member of the club and each edge represents a tie between two members of the club. It represents the presence (1) or absence (0) of ties among the members of the club. Hence, the work is carried out to predict the indirect ties that exist between the members of the club in future. From this, the friendships of the karate club in future will be predicted.

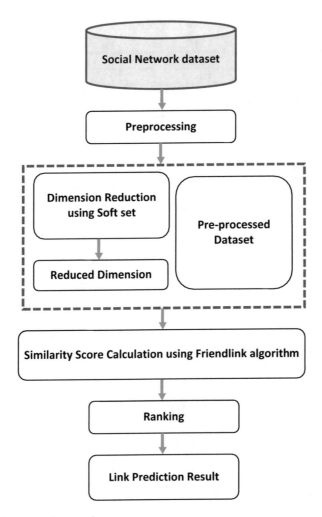

Fig. 17.2 The proposed approach

For finding the strong friendship between the nodes, the similarity score between two nodes before dimensionality reduction (**FLP**) is calculated using Friendlink algorithm. It considers all indirect links of length-2 and length-3 path. Out of all the nodes, some nodes are extracted and are shown in the Table 17.1.

In this table the zero represents that there is no link between the nodes with the path length of length-2 path and length-3 path and a nonzero value represents that the nodes are having a probability of some value to get linked in future. For example, the $node_1$ is having indirect link with many other nodes like $node_{24}$ and $node_{26}$. But it is found that $node_1$ is having highest similarity score with $node_{26}$ when compared to all other remaining nodes. It is concluded that $node_1$ will have higher probability to link with $node_{26}$ in future. The graphical representation for the values of

Table 17.1 Similarity scores of Friendlink Link Prediction

Nodes	9	24	26	32
1	0	0.000504	0.031754	0
2	0.03125	0	0	0
3	0	0.063004	0.000504	0.03125
4	0	0	0	0
5	0	0	0	0
6	0	0	0	0
7	0	0	0	0
8	0	0	0	0
9	0	0.03125	0.001008	0.063004
10	0.032258	0.001512	0.000504	0.031754
11	0	0	0	0
12	0	0	0	0
13	0	0	0	0
14	0.032258	0.001512	0.000504	0.031754
15	0.064012	0.03125	0.001008	0.063004
16	0.064012	0.03125	0.001008	0.063004
17	0	0	0	0
18	0	0	0	0
19	0.064012	0.03125	0.001008	0.063004
20	0.032258	0.001512	0.000504	0.031754
21	0.064012	0.03125	0.001008	0.063004
22	0	0	0	0
23	0.064012	0.03125	0.001008	0.063004
24	0.031754	0	0	0
25	0.032258	0.0625	0	0.031754
26	0	0.000504	0	0
27	0	0.03125	0	0
28	0.032258	0.001512	0.000504	0.031754
29	0.032258	0	0.031754	0.031754
30	0.064012	0.03125	0.001008	0
31	0.064012	0.03125	0.001008	0.063004
32	0.064012	0.03125	0.001008	0.063004
33	0.032258	0.001512	0.000504	0.031754
34	0	0	0	0

$node_1$ against all other remaining nodes is shown in Fig. 17.3a. Similarly, the $node_{15}$ is having links with other nodes such as $node_9$ and $node_{32}$, and it is found that $node_9$ is having higher similarity score. Therefore $node_{15}$ is having higher probability to be linked with $node_9$. The graphical representation for the values of $node_{15}$ against all other remaining nodes is shown in Fig. 17.3b.

Since the data set is immense, the dimensionality reduction technique is applied for extracting the friendship without affecting the members in the karate club. The

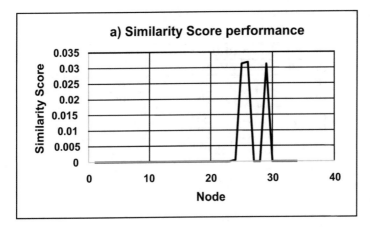

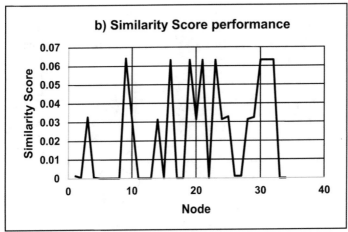

Fig. 17.3 Similarity scores of Friendlink Link Prediction

similarity score after applying **S**oft set-based **FLP** (**SFLP**) algorithm is calculated and is shown in Table 17.2.

It also considers all indirect links of length-2 and length-3 path. For example, with reduced dimension, the $node_1$ is having indirect link with many other nodes such as $node_{26}$ and $node_{30}$. But it is found that $node_1$ is having highest similarity score with $node_{26}$ with respect to all other remaining nodes even after the dimensionality reduction technique is employed. So, it is concluded that the $node_1$ will have a chance to be linked with $node_{26}$. The graphical representation for the values of $node_1$ against all other remaining nodes is shown in Fig. 17.4a. Similarly, the $node_{15}$ is having links with other nodes such as $node_9$ and $node_{26}$, and it is found that $node_9$ is having higher similarity score. Therefore $node_{15}$ is having higher probability to be linked with $node_9$. Hence it is observed that the node that is having an indirect link of length-2 and length-3 path length will be similar even after the

Table 17.2 Similarity scores of Soft set-based Friendlink Link Prediction

Nodes	9	26	30	31
1	0	0.094254	0.09375	0.001008
2	0.000504	0.0625	0	0
3	0	0	0.064012	0.032258
4	0.001008	0.03125	0.03125	0.001008
5	0	0.000504	0	0
6	0	0.001512	0	0
7	0	0.001008	0	0
8	0	0	0	0
9	0.03125	0.032762	0.03125	0
10	0.03125	0.001008	0.001512	0.031754
11	0	0	0	0
12	0	0	0	0
13	0	0	0	0
14	0	0.001008	0.001512	0.031754
15	0.063508	0.032762	0.001008	0.03125
16	0.03125	0.032762	0.001008	0.063508
17	0	0	0	0
18	0	0	0	0
19	0.03125	0.032762	0.001008	0.063508
20	0.03125	0.001008	0.001512	0.031754
21	0.03125	0.032762	0.001008	0.063508
22	0	0	0	0
23	0.03125	0.032762	0.001008	0.063508
24	0	0	0	0.031754
25	0.03125	0	0.001512	0.031754
26	0	0	0	0
27	0	0	0	0
28	0.03125	0.001008	0.001512	0.031754
29	0.03125	0.001008	0.001512	0.031754
30	0.03125	0.032762	0.001008	0.063508
31	0.03125	0.032762	0.001008	0.063508
32	0.03125	0.032762	0.001008	0.063508
33	0.03125	0.001008	0.001512	0.031754
34	0	0	0	0

dimensionality reduction. The graphical representation for the values of $node_{15}$ against all other remaining nodes is shown in Fig. 17.4b.

So, it is observed that the node that will occur in future is same for both the algorithms. Finally the consistency of the data set is preserved. The running time for executing the FLP algorithm is 0.1525 s and SFLP is 0.0147 s. It is found that the approach with dimensionality reduction technique works significantly better.

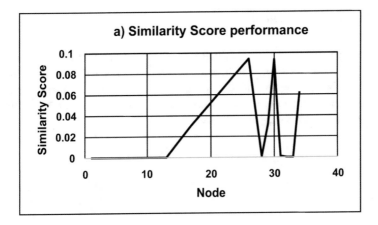

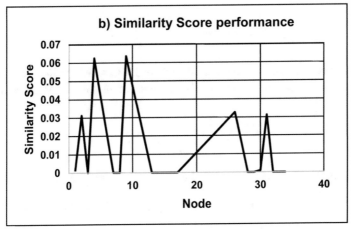

Fig. 17.4 Similarity scores of Soft set Friendlink Link Prediction

17.7 Centrality Measures

In SNA, the graph theory is exceptionally essential to discover the "central" actors. The graph theoretic thoughts measure an individual actor's prominence quality depending on centrality and prestige. An actor is said to be prominent if the ties of the actor make the actor particularly noticeable to the other actors in the network. Centrality is a measure of the importance of a node in a network. A prestigious actor is characterized as one who is the object of broad ties, thus concentrating exclusively on the actor as a recipient. Obviously, prestige is a more refined idea than centrality and can't generally be estimated. So, the centrality measures are considered to be significant. The commonly used centrality measures are degree, closeness, betweenness, eigenvector centrality, and clustering coefficient.

17.7.1 Degree Centrality

The degree centrality can be of, in-degree centrality, and out-degree centrality. The in-degree centrality is defined as an actor who receives many ties, and considered as prominent. The out-degree centrality is defined as an actor who disperses the information quickly to many others in a network and is categorized as influential. The degree centrality index [25] is given by

$$C_D(n_i) = d(n_i) \tag{17.3}$$

where C_D is the degree centrality for the ith node denoted by n_i, and $d(n_i)$ is the degree of the node n_i. This measure focuses on the most visible actors in the network.

17.7.2 Closeness Centrality

The closeness centrality is a measure of the degree to which an individual is close to all other individuals in a network. The closeness centrality is computed by using

$$C_C(n_i) = \left[\sum_{j=1}^{g} d(n_i, n_j) \right]^{-1} \tag{17.4}$$

where C_C is the closeness centrality for the ith node denoted by n_i, $d(n_i, n_j)$ is the length of any shortest path from n_i to n_j, $\sum_{j=1}^{g} d(n_i, n_j)$ is the total distance from all other nodes to node i. This measure depends on the whole of the geodesic distances from one another to all others. However, it is ambiguous for complicated graphs.

17.7.3 Betweenness Centrality

The betweenness centrality [26] is a measure of the extent to which a node is connected to other nodes that are not connected to each other. The betweenness centrality index is computed by using

$$C_B(n_i) = \sum_{j<k} g_{jk}(n_i) / g_{jk} \tag{17.5}$$

where C_B is the betweenness centrality for the ith node denoted by n_i , $g_{jk}(n_i)$ is the number of geodesics linking two nodes j and k that contain the node i. This measure estimates the degree of gatekeeping for every pair of actors in the network, concentrating on how much gatekeeping the subsequent actor accomplishes for the first.

17.8 Eigenvector Centrality

The eigenvector centrality is used to quantify the relative importance of a node in a network. The eigenvector centrality is calculated using

$$C_E\left(n_i\right) = \frac{1}{\lambda}\sum_{n=1}^{i} a_{ij} * x_j \tag{17.6}$$

where C_E is the eigenvector centrality for the ith node denoted by n_i, λ is a constant, a_{ij} is equal to 1 if the ith and j^{th} vertices are adjacent, that is, there is an edge connecting the ith and j^{th} vertices. a_{ij} is equal to 0 otherwise, that is, if there is no edge connecting the ith and j^{th} vertices.

17.8.1 Clustering Coefficient

The clustering coefficient is the measure of the degree to which a node in a network tends to cluster together. It is also known as local clustering coefficient. The clustering coefficient is calculated using

$$Cu_{Co}\left(ni\right) = \frac{\text{number of pairs of neighbors connected by edges}}{\text{number of pairs of neighbors}} \tag{17.7}$$

where Cu_{Co} is the clustering coefficient for the ith node denoted by n_i. This measure has become popular in *Nature* journal by Watts and Strogatz [27]. The centrality measures are computed using the open source tools for the original data set and is tabulated in Table 17.3.

It is observed that $node_{34}$ and $node_1$ are having high degree centrality and is considered as the prominent as well the most influential node in the Zachary's Karate Club data set. The $node_1$ has high betweenness centrality which means that through $node_1$ only most of the nodes are connected to other nodes in a network. The $node_1$ will be the gatekeeper for every pair of nodes in a network. The degree centrality of $node_1$ is same as that of the node that is having high betweenness centrality. So, the degree centrality is directly proportional to the betweenness centrality.

The closeness centrality for the nodes such as $node_{15}$ and $node_{23}$ is higher, thereby indicating that those nodes are very close to all other nodes in a network. The $node_{34}$ is having higher eigenvector value which shows that the node is relatively important

Table 17.3 Centrality measure calculation of Friendlink Link Prediction

Vertex	Degree Centrality	Betweenness Centrality	Closeness Centrality	Eigenvector Centrality	Clustering Coefficient
1	8	0.852	2.200	0.221	0.143
2	5	0.042	2.467	0.165	0.300
3	8	0.423	2.033	0.277	0.143
4	5	0.030	2.600	0.143	0.350
5	1	0.000	3.933	0.007	0.000
6	2	0.000	3.900	0.008	0.500
7	4	0.523	2.967	0.041	0.083
8	4	0.000	2.633	0.134	0.500
9	5	0.249	2.033	0.263	0.250
10	1	0.000	2.800	0.076	0.000
13	2	0.000	3.133	0.060	0.500
14	1	0.000	2.800	0.076	0.000
15	2	0.000	2.600	0.144	0.500
16	2	0.000	2.600	0.144	0.500
17	2	0.000	3.900	0.008	0.500
19	2	0.000	2.600	0.144	0.500
20	1	0.000	2.800	0.076	0.000
21	2	0.000	2.600	0.144	0.500
23	2	0.000	2.600	0.144	0.500
24	4	0.191	2.600	0.128	0.083
25	4	0.048	2.300	0.156	0.250
26	2	0.000	2.867	0.067	0.500
27	1	0.000	3.433	0.028	0.000
28	4	0.113	2.167	0.169	0.083
29	3	0.017	2.300	0.162	0.167
30	4	0.201	2.467	0.170	0.167
31	4	0.073	2.233	0.215	0.250
32	6	0.421	1.967	0.245	0.133
33	12	0.708	1.867	0.412	0.091
34	16	1.000	1.833	0.456	0.054

Table 17.4 Topological properties of the real data set for Friendlink Link Prediction

UE	EWD	TE	SL	V	GD
78	0	78	0	34	0.069518

with respect to other nodes in a network. The clustering coefficient for some of the nodes like $node_{15}$, $node_8$, $node_{19}$ that have higher values shows the property of one's friends are also friends of each other. Also, degree centrality is inversely proportional to the closeness centrality. It is concluded that nodes having low degree and betweenness centrality can be removed. The density of the graph G is the ratio of

Table 17.5 Centrality measure calculation of Soft set Friendlink Link Prediction

Vertex	Degree Centrality	Betweenness Centrality	Closeness Centrality	Eigenvector Centrality	Clustering Coefficient
1	16	1.000	1.758	0.356	0.075
2	9	0.123	2.061	0.266	0.167
3	10	0.328	1.788	0.317	0.122
4	6	0.027	2.152	0.211	0.333
5	3	0.001	2.636	0.076	0.333
6	4	0.069	2.606	0.079	0.250
7	4	0.069	2.606	0.079	0.250
8	4	0.000	2.273	0.171	0.500
9	5	0.128	1.939	0.227	0.250
10	2	0.002	2.303	0.103	0.000
11	3	0.001	2.636	0.076	0.333
12	1	0.000	2.727	0.053	0.000
13	2	0.000	2.697	0.084	0.500
14	5	0.105	1.939	0.226	0.300
15	2	0.000	2.697	0.101	0.500
16	2	0.000	2.697	0.101	0.500
17	2	0.000	3.515	0.024	0.500
18	2	0.000	2.667	0.092	0.500
19	2	0.000	2.697	0.101	0.500
20	3	0.074	2.000	0.148	0.167
21	2	0.000	2.697	0.101	0.500
22	2	0.000	2.667	0.092	0.500
23	2	0.000	2.697	0.101	0.500
24	5	0.040	2.545	0.150	0.200
25	3	0.005	2.667	0.057	0.167
26	3	0.009	2.667	0.059	0.167
27	2	0.000	2.758	0.076	0.500
28	4	0.051	2.182	0.133	0.083
29	3	0.004	2.212	0.131	0.167
30	4	0.007	2.606	0.135	0.333
31	4	0.033	2.182	0.175	0.250
32	6	0.316	1.848	0.191	0.100
33	12	0.332	1.939	0.309	0.098
34	17	0.695	1.818	0.374	0.055

edges in *G* to the maximum number of edges. The graph density is computed for the original data set and shown in Table 17.4. The topological properties are Unique Edges (UE), Edges With Duplicates (EWD), Total Edges (TE), Self-Loops (SL), Vertices (V), and Graph Density (GD).

The centrality measures are computed for the data set for which the soft set theory is applied and finally the reduced dimension data set is obtained. The centrality measure is tabulated in Table 17.5.

Table 17.6 Topological properties of the real data set for Soft set Friendlink Link Prediction

UE	EWD	TE	SL	V	GD
60	0	60	0	34	0.064516

The degree for $node_{34}$ and $node_1$ is higher, so the original data set $node_{34}$ and $node_1$ are prominent and influential nodes. The betweenness centrality for the $node_1$ is higher which is same as that of the centrality measures obtained for the original data set. The remaining metrics are also the same. Hence, the centrality measures for both original data set and reduced dimension are compared. It is agreed that the node having low degree and betweenness centrality is removed without affecting the consistency of the data set, so it improves the memory space utilization. The graph density is computed and shown in Table 17.6.

It is observed that the graph density is compressed for the reduced dimension.

17.9 Theoretical Analysis

The proposed work comprises of three tasks, the first task is to construct a reduced data set and the time complexity for constructing is $O(n + e)$, where n is the nodes and e is the number of edges, always $n < e$. The second task is to find similarity score using Friendlink algorithm and the time complexity is $O(n * a)$, where n is the number of nodes and a is the average degree of the network. So, the time complexity of these two tasks would be $O(\max(n + e, n * a))$, since the average degree is less than the number of edges, the time complexity would be $O(n + e)$. The third task is to rank all the similarity scores and the time complexity is $O(n \log n)$. Hence, the total time complexity of the proposed algorithm would be $O(\max(n + e, n \log n))$, since the number of nodes is comparatively less than the number of edges, the time complexity would be $O(n \log n)$ and the space complexity would be $O(n_1 * a_1)$, where a_1 is the average node degree and n_1 is number of nodes after dimensionality reduction which is strictly less than the number of nodes present in the original data set. The proposed technique significantly improves the performance of link prediction compared to Katz [28] index, since it needs the inverse matrix representation for further computation. For implementation, the adjacent list representation is used for storing the adjacent nodes in a graph.

17.10 Empirical Analysis

Most standard evaluation metric is Mean Average Precision (MAP), which provides a single-figure quality measure. It is defined by

$$MAP = \sum_{n=1}^{i} AveP(n) \qquad (17.8)$$

where n is the total number of nodes, i denotes the iteration, $AveP(n)$ is the average precision value. It has been shown to have especially good discrimination and stability among the various metrics. So, the link prediction accuracy is evaluated using MAP for both algorithms. The MAP value for SFLP is 0.38853 and FLP is 0.07865. The empirical analysis shows that MAP value is higher for SFLP than FLP algorithm.

17.11 Conclusion and Future Work

Social networks play a significant role in all domains, in which the data set is very large. This chapter concentrates on soft set theory that has been applied for dimensionality reduction. The Friendlink algorithm is used for finding the similarity score of a node with other nodes. Both the Soft set theory and Friendlink algorithm are applied for the real data set in order to predict the node that will occur in future. Then the consistency of the data is verified against the various social network metrics. The experimental results show that the approach works better in terms of space. Further improvement might be predicting the links that can be applied to various domains like community detection, name disambiguation, and health care.

References

1. Wasserman, S., & Faust, K. (1994). *Social network analysis: Methods and applications* (Vol. 8). Cambridge: Cambridge University Press.
2. Song, Han Hee, Tae Won Cho, Vacha Dave, Yin Zhang, and Lili Qiu. (2009) Scalable proximity estimation and link prediction in online social networks. In *Proceedings of the 9th ACM SIGCOMM conference on Internet measurement conference, ACM* (pp. 322–335).
3. Aldous, J. M., & Wilson, R. J.. (2003) *Graphs and applications: An introductory approach* (Vol. 1). Springer.
4. Liben-Nowell, D., & Kleinberg, J. (2007). The link-prediction problem for social networks. *Journal of the Association for Information Science and Technology, 58*(7), 1019–1031.
5. Chen, K.-J., Han, J., & Li, Y. (2014). HALLP: A hybrid active learning approach to link prediction task. *JCP, 9*(3), 551–556.
6. Sharma, D., Sharma, U., & Khatr, S. K. (2014). An experimental comparison of the link prediction techniques in social networks. *International Journal of Modeling and Optimization, 4*(1), 21.
7. Michael, F., Tenenboim-Chekina, L., Puzis, R., Lesser, O., Rokach, L., & Elovici, Y. (2013). Computationally efficient link prediction in a variety of social networks. *ACM Transactions on Intelligent Systems and Technology (TIST), 5*(1), 10.
8. Gupta, N., & Singh, A.. (2014). A novel strategy for link prediction in social networks. In *Proceedings of the 2014 CoNEXT on Student Workshop, ACM* (pp. 12–14).

9. Kurakar, M., & Izudheen, S. (2014). Link prediction in protein-protein networks. *Survey: International Journal of Computer Trends & Technology, 1*(9), 164–168.

10. Almansoori, Wadhah, Shang Gao, Tamer M. Jarada, Reda Alhajj, & Jon Rokne. (2011) Link prediction and classification in social networks and its application in healthcare. In Information Reuse and Integration (IRI), 2011 IEEE International Conference on, IEEE, pp. 12–14.

11. Papadimitriou, A., Symeonidis, P., & Manolopoulos, Y. (2012). Fast and accurate link prediction in social networking systems. *Journal of Systems and Software, 85*(9), 2119–2132.

12. Gong, N. Z., Talwalkar, A., Mackey, L., Huang, L., Shin, E. C. R., Stefanov, E., & Song, D. (2012). Jointly predicting links and inferring attributes using a social-attribute network (san). *ACM*, 1–9.

13. Menon, A. K., & Elkan, C. (2011). Link prediction via matrix factorization. In *Joint European conference on machine learning and knowledge discovery in databases* (pp. 437–452). Berlin Heidelberg: Springer.

14. Dunlavy, D. M., Kolda, T. G., & Acar, E. (2011). Temporal link prediction using matrix and tensor factorizations. *ACM Transactions on Knowledge Discovery from Data (TKDD), 5*(2), 10.

15. Zhao, J., Xu, F., Dong, L., Liang, X., & Xu, K. (2012). Performance of local information-based link prediction: A sampling perspective. *Journal of Physics A Mathematical and Theoretical, 45*(34), 345001.

16. Lichtenwalter, R. N., Lussier, J. T., & Chawla, N. V. (2010). New perspectives and methods in link prediction. *Proceedings of the 16th ACM SIGKDD international conference on Knowledge discovery and data mining ACM*, 243–252.

17. Yu, G. (2015). On the parameter reduction of soft binary relations. *Annals of Fuzzy Athematics and Informatics*, 1–10.

18. Ma, X., & Qin, H. (2014). Application of a new efficient Normal parameter reduction algorithm of soft sets in online shopping. *World Academy of Science, Engineering and Technology, International Journal of Computer, Electrical, Automation, Control and Information Engineering, 8*(7), 1174–1176.

19. Liben-Nowell, D., & Kleinberg, J. The link-prediction problem for social networks. *Journal of the American Society for Information Science and Technology, 58*(7), 1019–1031.

20. Yu, C., Zhao, X., Lu, A., & Lin, X. (2017). Similarity-based link prediction in social networks: A path and node combined approach. *Journal of Information Science, 43*(5), 683–695.

21. Molodtsov, D. (1999). Soft set theory—First results. *Computers & Mathematics with Applications, 37*(4–5), 19–31.

22. Zimmermann, H.-J. (2011). *Fuzzy set theory—And its applications.* Springer.

23. Papadimitriou, A., Symeonidis, P., & Manolopoulos, Y. (2011). Friendlink: link prediction in social networks via bounded local path traversal. In *Computational Aspects of Social Networks (CASoN)* (pp. 66–71). 2011 International Conference on, IEEE.

24. Rubin, F. R. A. N. K. (1978). Enumerating all simple paths in a graph. *IEEE Transactions on Circuits and Systems, 25*(8), 641–642.

25. Baagyere, E. Y., Qin, Z., Xiong, H., & Zhiguang, Q. (2016). The structural properties of online social networks and their application areas. *IAENG International Journal of Computer Science, 43*(2).

26. Freeman, L. C. (1977) *A set of measures of centrality based on: (1977): 35- betweenness.* Sociometry 41.

27. Watts, D. J., & Strogatz, S. H. (1998). Collective dynamics of 'small-world' networks. *Nature, 393*(6684), 440.

28. Katz, L. (1953). A new status index derived from sociometric analysis. *Psychometrika, 18*(1), 39–43.

Chapter 18
Overview on Deep Neural Networks: Architecture, Application and Rising Analysis Trends

V. Niranjani and N. Saravana Selvam

18.1 Introduction

Alongside Big Data [1], the subject of deep education has become a dominant subject for the industry and research fields for the growth of different intelligent systems, Cloud Computing [2] and the Internet of Things (IoT) [3]. The ability to estimate and reduce comprehensive, complicated data sets into extremely precise, transformative output has demonstrated an important potential [4, 5]. Conversely, deep neural networks (DNN) architectures may be applied to all kinds of information—whether numerical, visual, texts, audio, or a certain mixture—as opposed to complicated hard-coded programs for one inflexible job only. In addition, sophisticated, often open-source and widely accessible advanced, deep learning platforms become increasingly accessible. Moreover, significant businesses such as Amazon, Flipkart, Microsoft, Apple, and others are investing strongly in profound learning technology.

Supervised classification tasks have outstripped human capacities in fields like manuscript and picture recognition, one in each of the key fields for deep learning applications [6]. Furthermore, unattended data set learning without specific labels have shown the possibility of extraction by means of the clustering and statistical analysis of unpredicted assessment and business value. Most potentially attractive yet is strengthened learning, providing feedback from a linked setting that offers the opportunity for deep learning without human oversight. The field of robotics and computer view was widely used for this sort of profound learning.

V. Niranjani (✉)
Sri Eshwar College of Engineering, Coimbatore, India

N. S. Selvam
PSR Engineering College, Sivakasi, India

© Springer Nature Switzerland AG 2020   271
A. Haldorai et al. (eds.), *Business Intelligence for Enterprise Internet of Things*,
EAI/Springer Innovations in Communication and Computing,
https://doi.org/10.1007/978-3-030-44407-5_18

As IoT and smart-world systems are steadily growing, powered by CPS (cyber-physical system) technologies in which networking in all devices and capable of communicated sensitive information and physical items monitor, bigger and bigger datasets are accessible for profound learning which can have a material effect on our everyday life [6, 7–9]. For example, smart phones have provided the ability to create rich video, audio, and text data through integrated GPS modules from various media apps and integrated sensors, as well as massively locating population movement. All such apps are clearly generated, alone, or together, by unprecedented Big Data. Intelligent cities are designed to allow resource management in almost all fields like communications, electricity, transportation, emergency, and other utilities to building. Intelligent transport systems will interconnect auto-drive cars and infrastructure networks to revolutionize day-to-day mass transit, eliminate accidents, and facilitate the storage of the secondary grid.

18.2   Overview of Deep Neural Networks

Machine learning (ML) includes a wide range of algorithms, which cannot all be categorized as deep learning. Specific algorithms like Bayesian algorithms are restricted in implementation and in the capacity to learn massive complicated information representations, including statistical processes such as linear-regression or decision-making trees. The artificial intelligence field is primarily when machines can perform functions that typically require human intelligence [10]. It includes machine learning, where machines can obtain abilities through experience and without participation from human beings [11]. Deep learning is a component of machine learning that is used to learn from big quantities of information by artificial neural networks, human brain-inspired algorithms. Similar to how we learn from experience, each time the deep learning algorithm tweaks a job a bit in order to enhance the result. We are talking about 'profound learning' because neural networks have different (profound) levels that allow learning to take place [12].

The quantity of information we produce each day—currently estimated at 2.6 quintillion octets—is astonishing and is the resource that enables profound learning. Since profound learning algorithms require a ton of information to draw on, this rise in information development is one reason why in the latest years profound learner skills have grown. In relation to further development of information and the proliferation of artificial intelligence (AI), deep learning algorithms profit from today's greater computing power. AI as the service has offered smaller organizations access, especially the AI algorithms needed for deep learning without a big one, to artificial intelligence technology shown in Fig. 18.1.

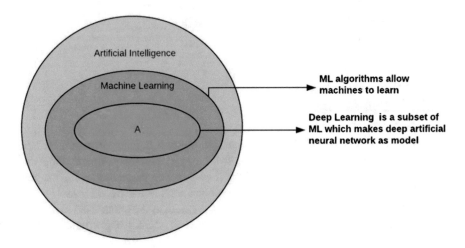

Fig. 18.1 Diagram of ML vs. DL

18.2.1 Deep-Neural Network Architecture

Three significant layers are input layer, hidden layers, and output layers for deep-learning architectures. The number of layers concealed determines the architectural depth. Various nonlinear features can be learned depending on the sort of concealed layers used. Equations (18.1) and (18.2) are straightforward ANN patterns. In Eq. (18.1) the nonlinearity observed within the information (concealed layer) is where the activation function determines the nonlinearity features of the model. The same applies to Eq. (18.2) (Output layer) which translates the feature of nonlinearity into a forecast. An easy model for deep learning has several hidden layers.

$$AC_1 = Act\left(Z1A + C_1\right)$$ (18.1)

$$\hat{Y} = Z_2 AC_1 + C_2$$ (18.2)

18.2.2 Activation Function

The hidden layer activation function helps to map the nonlinearity connection between input and output. In concealed layers, most used activation functions are sigmoid and hyperbolic (Tanh). Specific activation functions are not applicable. For datasets, various activation functions must be assessed. In this document, we use the rectified linear activation function (also known as "relu") (see Eq. (18.3)), since the techniques of the gradation optimization facilitate model training [13].

$$f(x) = \begin{cases} 0, & x < 0 \\ x, & x \geq 0 \end{cases}$$

(18.3)

18.2.3 Hidden Layer

Equation (18.1) is a straightforward concealed layer. A number of parameters are used to define the complexity of each hidden layer. The number of parameters that are the number of concealed units and the regularization parameter in L2 are governed by hyper parameters. The number of hidden parameters in each layer (i.e., the weights) and the regularization of L2 reduce the extent of the parameters in order to avoid over fitting [13]. To ensure a good model fit, tuning hyper parameters is essential.

18.2.4 Outer Layer

Typically, the output comprises inactivation of an identity (also known as linear activation) for regression issues. Activation of identity allows for adverse projections. As the demand for cooling and heating energy is non-negative reaction values with a minimum "zero," identification and activation for this implementation is not appropriate. The linear activation function is used rather than rectified in Fig. 18.2.

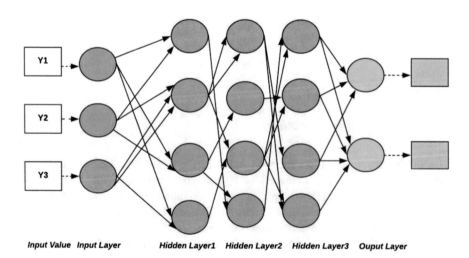

Fig. 18.2 Diagram of deep learning architecture

18.3 Deep-Neural Network Architecture

18.3.1 Supervised Learning

Supervised learning (SL) is named because of the need for consistent marking of the studied data and for tracking or classifying the study results as accurate or incorrect. Supervised learning(SL) is, in particular, used as a mechanism of prediction where some of the data are learned (also known as training-set), a second part to validate a model which are already trained (cross-validation), and the remaining data are used to the predict the accuracy and effectiveness. While accuracy is a significant measure, the ability of a trained model to generalize to new information is used through other statistically significant processes, such as precision, recall, and F1 scores. Classification and regression are the two main tasks for monitored learning.

18.3.2 Unsupervised Learning

In unsupervised learning, datasets are not marked in any manner that identifies a right or improper outcome, given as input for machine learning. Rather, the outcome may reach a larger required objective, the capacity to find something easy to understand by humans or a complicated use of a statistical function to obtain the required value may be assessed. For example, the clustering algorithm may cluster-data or fugitive groups, but it cannot be simple to tell if the clustering is actually correct without an appropriate visual representation. Similarly, the density estimate provides only for an estimate to be used for compression or a decrease in dimension that is either relevant or irrelevant to a dataset or to encode information effectively. However, it may still be necessary to find the extraction ability of compressed images accurately to establish the adequacy of implementation.

18.3.3 Reinforcement Learning

Reinforcement Learning is an intermediate between monitored and uncontrolled learning, although information is not specifically labeled, there is a reward for the performance in each action. In particular, the architecture of this learning is enhanced learning that interacts directly with the setting, thus providing a particular reward to a changing setting. In order to enlarge the benefit to all the state transitions, the enhanced education scheme aims to learn the best measures to be taken at every state. This can be done for an endless moment or can be implemented in meetings to maximize the results of each session through a perception-action-learning loop. Furthermore, the feedback can either be provided straight from the setting or,

Table 18.1 Different categorization of deep learning

	Supervised learning	Unsupervised learning	Reinforcement learning
Definition	Both predictors and predictions exist for the training set.	Training package has only data set predictors.	In every assignment they can produce cutting-edge outcomes.
Algorithm	SVM, Linear regression and logistic regression, Naive Bayes	Dimensionality reduction, K-Means, Clustering	Q-Learning, SARSA, DQN
Uses	Forecasting	Pre-process the data, pre-train supervised learning algorithms.	Warehouses, Inventory management, delivery management, Power system, Financial systems.

as a consequence of some calculation or function, by way of numerical counter in an interactive setting shown Table 18.1.

The search policy and approximation function can be divided between the two main means of enhancement. Policy search can be performed using gradient-based (back propagation) or gradient-free (evolutionary) techniques to search for an ideal policy directly. Value function is a technique through which an evaluation of the expected return of a particular state and the selection of an ideal strategy to select an action to maximize the desired value are required for each state action.

18.4   Applications of Deep Neural Networks

Here we explore the primary applications of deep learning. An important body of job has continuously evolved over the past few years towards the implementation of deep learning. In particular, the main advances were in applying deep neural networks to analyze the multimedia data including image, video, audio, and NLP that resulted in significant state-of-the-art leaps for all systems. In fact, ML is primarily worried about information fitting, whose main uses are discrimination, prediction, and optimization. In addition, the progress made in Big Data and the Cloud Computing sector has created a chance for machine learning to flourish, which enables the information collection, dissemination, and computational model performances. The presence of the information and the nature of their potential immediately required more precise, widespread, and effective processes of learning.

18.4.1   Fraud Detection

The financial and banking industry is an area that benefits from deep learning and is affected by the job of detecting fraud with digital cash transactions. Fraud detection is performed on the basis of identification of patterns and credit scores for client operations, anomalous conduct, and outlines. Auto-encoders are created for credit-card-fraud in Tensor flow and Keras that save billions of USD of financial institution recapture and insurance price. Machine learning methods and neural networks are used for fraud detection. Mainly machine learning is used to highlight cases of fraud which require human deliberation; deep learning is intended to minimize such efforts.

18.4.2   Diagnostics Medical

Highly affected by advances in image analysis, the fast improvements in deep learning have greatly benefited medical diagnostics. Considerable research has been performed to improve the detection of pictures of MRI, tumors, CT scans, illnesses, and other abnormalities. Furthermore, the devices of IoT for medical application are providing independent patient surveillance and to extract helpful medical population information.

18.4.3   Self-Driving Car

Deep learning gives life to self-employed movement. One million set of information are supplied in a scheme for model building, machine training, and ensuring that outcomes are tested in secure setting. Autonomous car developers are concerned with handling scenarios that are unprecedented. Typical to deep learning algorithms, a periodic cycle of testing and execution ensures secure driving and increasing exposure to millions of situations. Dash cams, geo-mapping, and sensor data help to generate brief and advanced models for navigating through traffic, identifying paths, signage, pedestrian-only routings, etc.

18.5   Summary

Deep learning is a technology that continues to evolve and has been implemented to great effect in a multitude of applications and domains. Deep learning methods are practical for us to fix a lot of issues. Although full-scale implementation of DL technology in industry is ongoing, calculated steps must be taken to ensure proper application of deep learning, as the subversion of deep learning models can lead to significant loss of financial value, confidence, or even extreme existence. We analyzed in detail deep learning architectures based on learning mechanisms (supervised, unsupervised, and reinforced) and target output models. In addition, in deep learning science, we have thoroughly investigated the state of the art. Such areas include the processing of multimedia (text, audio, and video), automated systems, medical diagnostics, financial applications, and security analysis. We hope that this work offers a useful reference for both scientists and computer science professionals in considering deep learning methods, and application, and causes interest in fields that urgently need further consideration.

References

1. Liang, F., Yu, W., An, D., Yang, Q., Fu, X., & Zhao, W. (2018). A survey on big data market: Pricing, trading and protection. *IEEE Access, 6,* 15132–15154.
2. Huang, H. H., & Liu, H. (2014, October). Big data machine learning and graph analytics: Current state and future challenges. In *Proceedings of IEEE international conference on Big Data (Big Data)* (pp. 16–17).
3. Yu, W., et al. (2017). A survey on the edge computing for the internet of things. *IEEE Access, 6,* 6900–6919.
4. Chen, X. W., & Lin, X. (2014). Big data deep learning: Challenges and perspectives. *IEEE Access, 2,* 514–525.
5. Nguyen, N. D., Nguyen, T., & Nahavandi, S. (2017). System design perspective for human-level agents using deep reinforcement learning: A survey. *IEEE Access, 5,* 27091–27102.
6. Nielsen, M.. (2017). *Neural networks and deep learning.* [Online]. Available: http://neuralnet-worksanddeeplearning.com/.
7. Xu, H., Lin, J., & Yu, W. (2017). *Smart transportation systems: Architecture, enabling technologies, and open issues* (pp. 23–49). Singapore: Springer.
8. Wu, J., & Zhao, W. (2016, November). Design and realization of internet: From net of things to internet of things. *ACM Transactions on Cyber-Physical Systems, 1*(1):1–2.
9. Ekedebe, N., Lu, C., & Yu, W. (2015, June). Towards experimental evaluation of intelligent transportation system safety and traffic efficiency. In *Proceedings of International Conference on Communications (ICC)* (pp. 3757–3762).
10. Shi, W., Cao, J., Zhang, Q., Li, Y., & Xu, L. (2016, October). Edge computing: Vision and challenges. *IEEE Internet of Things Journal, 3*(5): 637–646.
11. Arulkumaran, K., Deisenroth, M. P., Brundage, M., & Bharath, A. A. (2017, November). Deep reinforcement learning: A brief survey. *IEEE Signal Processing Magazine, 34,* no. 6, pp. 26–38.
12. Fadlullah, Z. M. et al. (2017). State-of-the-art deep learning: Evolving machine intelligence toward tomorrow's intelligent network traffic control systems. *IEEE Communications Surveys and Tutorials, 19*(4): 2432–2455, 4th Quart.
13. Goodfellow, I., Bengio, Y., Courville, A. (2016). *Deep Learn.* Google Scholar

Index

© Springer Nature Switzerland AG 2020

A. Haldorai et al. (eds.), *Business Intelligence for Enterprise Internet of Things*,
EAI/Springer Innovations in Communication and Computing,
https://doi.org/10.1007/978-3-030-44407-5

Printed in the United States
by Baker & Taylor Publisher Services